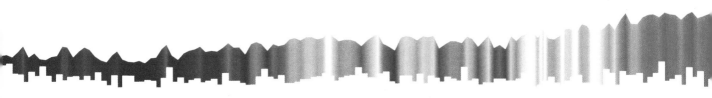

GLOBAL WARMING SCIENCE

GLOBAL WARMING SCIENCE

A Quantitative Introduction to Climate Change and Its Consequences

Eli Tziperman

PRINCETON UNIVERSITY PRESS

Princeton and Oxford

Published by Princeton University Press
41 William Street, Princeton, New Jersey 08540
6 Oxford Street, Woodstock, Oxfordshire OX20 1TR

press.princeton.edu

All Rights Reserved

Library of Congress Control Number: 2021945685
ISBN 9780691228808
ISBN (pbk.) 9780691228792
ISBN (e-book) 9780691228815

British Library Cataloging-in-Publication Data is available

Editorial: Ingrid Gnerlich, Whitney Rauenhorst
Jacket/Cover Design: Wanda España
Production: Jacqueline Poirier
Publicity: Matthew Taylor, Charlotte Coyne
Copyeditor: Elizabeth J. Asborno
Production Editor: Theresa Carcaldi (Westchester Publishing Services)

Jacket/Cover Credit: Images (clockwise from left): L'Albufera de Valencia, Spain. Courtesy of Wladimir Bulgar / Science Photo Library; Glacial lagoon, Iceland. Courtesy of Colin Bain / Alamy; Category 5 super typhoon, view from space. Courtesy of Evgeniy Baranov / Alamy; Coral reef at Palmyra Atoll. Courtesy of Jim Maragos / U.S. Fish and Wildlife Service. (center) Forest fire in Western Alberta, Canada. Courtesy of Cameron Strandberg

This book has been composed in Arno and AvenirNext

Printed on acid-free paper. ∞

Printed in the United States of America

10 9 8 7 6 5 4 3 2 1

CONTENTS

Preface   xi

1   OVERVIEW   1

1.1   Workshop   10

2   GREENHOUSE   11

2.1   The greenhouse effect   13

  2.1.1   Earth's energy balance   13

  2.1.2   The greenhouse effect: a two-layer model   14

  2.1.3   The emission height and lapse rate   17

2.2   Greenhouse gases   19

  2.2.1   Wavelength-dependent black-body radiation   19

  2.2.2   Energy levels and absorption   20

  2.2.3   Broadening   21

  2.2.4   Radiative forcing, logarithmic dependence on CO_2 22

  2.2.5   Other greenhouse gases, global warming potential   24

  Box 2.1: The Clausius-Clapeyron relation   25

  2.2.6   The water vapor feedback   26

2.3   Workshop   28

3	**TEMPERATURE**	**31**
3.1	Climate sensitivity and the role of the ocean	34
	3.1.1 Equilibrium climate sensitivity	34
	3.1.2 Transient climate sensitivity	36
3.2	Polar amplification	39
3.3	"Hiatus" periods	44
3.4	Stratospheric cooling	45
	3.4.1 Detection and attribution	48
3.5	▄▄ Workshop	49
4	**SEA LEVEL**	**53**
4.1	Global mean sea level changes	56
	4.1.1 Thermal expansion	56
	▓▓ **Box 4.1:** Past warm climates	59
	4.1.2 Ice sheets and mountain glaciers	61
	4.1.3 Land water storage	62
	4.1.4 Detection of anthropogenic climate change in GMSL	62
4.2	Regional sea level changes	64
	4.2.1 Atmosphere-ocean interaction	64
	4.2.2 Land changes	70
	4.2.3 Gravitational effects: sea level fingerprints of melting	71
4.3	▄▄ Workshop	73
5	**OCEAN ACIDIFICATION**	**77**
5.1	Calcium carbonate ($CaCO_3$) dissolution	80
	▓▓ **Box 5.1:** The carbon cycle	81
5.2	The carbonate system	82
	5.2.1 Carbonate system equations	85
	5.2.2 Approximate solution of the carbonate system	87
5.3	Response to perturbations	90
	5.3.1 Response to increased atmospheric CO_2 concentration	90

	5.3.2	Response to warming	91
	5.3.3	Long-term decline of anthropogenic CO_2	92
5.4		▰ Workshop	95

6   OCEAN CIRCULATION   99

	▰ **Box 6.1:** Ocean temperature, salinity, and water masses	102
6.1	Observations and projections	103
6.2	The Stommel model	106
6.3	Multiple equilibria, tipping points, hysteresis	110
6.4	Keeping it simple	113
6.5	Consequences of AMOC collapse	115
6.6	The oceans and global warming	116
6.7	▰ Workshop	119

7   CLOUDS   121

7.1	Cloud fundamentals	123
7.2	Moist convection and cloud formation	125
7.3	Cloud microphysics	131
7.4	Cloud feedbacks and climate uncertainty	134
7.5	▰ Workshop	138

8   HURRICANES   141

8.1	Factors affecting hurricane magnitude	142
8.2	Potential intensity	145
	▰ **Box 8.1:** El Niño, La Niña	146
8.3	Observed changes to hurricane activity	152
8.4	▰ Workshop	155

9   ARCTIC SEA ICE   157

| 9.1 | Processes and feedbacks | 161 |
| 9.2 | Detection of climate change | 165 |

| 9.3 | Future projections | 167 |
| 9.4 | ▪ Workshop | 169 |

10   GREENLAND AND ANTARCTICA   171

10.1	Terminology	173
10.2	Processes	174
	10.2.1 Accumulation	174
	10.2.2 Surface melting and PDD	175
	10.2.3 Calving	177
	10.2.4 Ice flow	179
	10.2.5 Basal hydrology	181
	▪ **Box 10.1:** Ice ages	185
10.3	Observed trends and projections	186
10.4	▪ Workshop	190

11   MOUNTAIN GLACIERS   193

11.1	Observed retreat	196
11.2	Mountain glaciers as a climate indicator	198
	11.2.1 Reconstructing temperature from glacier extent	198
	11.2.2 Ice cores from mountain glaciers	202
11.3	Glacier dynamics	203
11.4	Mountain glacier retreat in perspective	206
11.5	▪ Workshop	208

12   DROUGHTS AND PRECIPITATION   211

12.1	Relevant processes and terms	214
12.2	Why droughts happen, climate teleconnections	214
	▪ **Box 12.1:** The Indian Ocean dipole	217
12.3	Detection of climate change	219
12.4	Observations, paleo proxy data	221
12.5	Example projections: Southwest United States and the Sahel	222

12.6	Understanding precipitation trends	224
	12.6.1 Hadley cell expansion and weakening	225
	12.6.2 "Wet getting wetter, dry getting drier" projections	228
	12.6.3 Precipitation extremes in a warmer climate	231
12.7	A bucket model for soil moisture	237
12.8	▮▮ Workshop	242

13	**HEAT WAVES**	**245**
13.1	Physical processes	247
13.2	Heat stress	251
13.3	Future projections	253
13.4	▮▮ Workshop	258

14	**FOREST FIRES**	**261**
14.1	Tools	264
14.2	Detection of burnt area due to ACC	265
14.3	Fires and natural climate variability	269
14.4	Observed global trends and future projections	273
14.5	▮▮ Workshop	276

▮▮	Notes	279
▮▮	Bibliography	293
▮▮	Index	305

PREFACE

The purpose of this book is to provide a quantitative, undergraduate-level survey of the science involved in the study of anthropogenic global warming and its consequences, from hurricanes and droughts to ocean acidification and forest fires, and more. Each chapter attempts to explain the physical or chemical mechanisms behind the observed or anticipated change, to demonstrate the statistical or other tools that are being used to understand and predict these changes, and to address the uncertainty involved in both existing observations and future projections. The students can therefore gain a detailed understanding of *what* is expected to happen, *why* this is the case, and the *level and sources of uncertainty* for each subject. The focus is on climate science; the closely related issues of energy and policy are not addressed. The emphasis on *climate change* means that the basic operations of the climate system are only discussed when they are related to an observed or anticipated change. Several such basic climate and paleoclimate science topics that are important for understanding climate change are explained in "climate background boxes" spread throughout the text. Those of us working in climate research may feel that it is important to cover the fundamentals of how the climate system operates before approaching the subject of climate change. However, many students are likely to be driven to understand issues related to climate change that are often in the news, and perhaps exposing them to the science behind the news may motivate them to then study other aspects of climate science.

Much of the study of climate change is based on large-scale complex Earth System Models that attempt to simulate the oceans, atmosphere, land surface, cryosphere, and biosphere. Yet this book is based on the belief that every one of the relevant subjects can be understood using a simpler framework, employing a fairly straightforward statistical analysis, a simple ordinary differential equation, or a set of basic chemical reactions. Some of the issues are then also demonstrated by analyzing the results of complex climate models. The use of basic math throughout the discussion means that the students are assumed to have taken elementary college-level calculus and are aware of rudimentary concepts in statistics, although no prior college-level science preparation is assumed. This book may be used in a course for science, technology, engineering, and mathematics (STEM) students of all disciplines, or any non-STEM students who are not intimidated by quantitative reasoning, starting with their second year of undergraduate education. Given that this is meant as an introduction for undergraduates, the description does not necessarily represent the state-of-the-art science, although it is mostly based on selected recent papers, as indicated in the notes. The notes also contain pointers and references to the sources of all data and other materials used, plus occasional brief mentions of more refined issues that are beyond the scope here, with relevant references. When future projections are analyzed, they are mostly based on the Representative Concentration Pathway (RCP) 8.5 high-emission warming scenario (acronyms are defined in the index). While one hopes the future greenhouse gas emissions used in this scenario are unrealistically severe in terms of future emissions, the resulting projections clearly show possible climate trends, demonstrate the relevant mechanisms, and allow us to cleanly differentiate signal from noise.

Each chapter is followed by a workshop that is based on a Jupyter python notebook and an accompanying small data set, which are used to guide the students in analyzing, plotting, and understanding problems relevant to the material. These notebooks can be used as an active learning element for in-class workshops, alternating with short half-hour segments of lecture time, with the students then completing the workshops after class as homework assignments. The workshops and data allow the students to reproduce nearly all analyses and figures appearing in the text and occasionally also involve some simple pen-and-paper derivations. Each workshop concludes with writing guidelines for a one-page essay about the corresponding subject advising policymakers (e.g., the president science adviser) about the issues involved. These essays allow the students to synthesize the material covered and grasp the big picture. The notebook exercises and corresponding data sets, as well as slides to be used for teaching a class based on these notes, can be found at https://press.princeton.edu/global-warming-science, and the solutions to these notebook workshops are available to lecturers upon request. While this book focuses exclusively on the science aspects of global warming, when used for a course the instructor can also include a class about critical reading of the popular press about climate change and another about how climate science informs (or does not inform) climate policy. The materials for these classes are also available at the above link.

Many thanks to the graduate students who contributed significantly to four chapters, writing the first draft and much of the code used for these chapters, and teaching me about these subjects in the process: Camille Hankel ("Greenhouse"), Minmin Fu ("Clouds"), Xiaoting Yang ("Sea Level") and Wanying Kang ("Ice Sheets"). I am grateful to Peter Huybers for co-teaching with me a reading course on climate change debates for quite a few years, which taught me much of what I know about the subject and led to this incarnation of the course on the science of global warming. Thanks to Brian Farrell for sharing his deep insights on many issues, including some covered herein. Thanks to the students who took Earth and Planetary Sciences 101 and to the teaching assistants for their comments and enthusiasm. I am very grateful to, and cannot thank enough, the many colleagues who have read parts of the text—sometimes multiple chapters; they were exceedingly generous with their time and provided wonderful, insightful feedback: Adam Sobel, Alex Robel, Brad Lipovsky, Brendan Rogers, Claudia Pasquero, Dan Schrag, Dan Yakir, Eli Galanti, Golan Bel, Hezi Gildor, Ian Eisenman, Ilan Koren, Itay Halevy, Isaac Held, Jochem Marotzke, Jonathan Gilligan, Laure Zanna, Mark Cane, Orit Altaratz, Park Williams, Paul

O'Gorman, Peter Huybers, Raffaele Ferrari, Rei Chemke, Robbie Toggweiler, Shira Raveh-Rubin, Steve Griffies, Yani Yuval, Yochanan Kushnir, Yongyun Hu, Yossi Ashkenazy, and Zhiming Kuang. Finally, many thanks to my friends and colleagues in the Earth and Planetary Science department at the Weizmann Institute for their wonderful hospitality and to those at Harvard for creating such a friendly and stimulating environment.

GLOBAL WARMING SCIENCE

OVERVIEW

Consider a brief overview of the issues to be surveyed in the following chapters. This overview outlines observed and expected changes, our ability to attribute observed trends and events to anthropogenic climate change, the level and reasons for the uncertainties involved in quantifying both observed and projected changes, and the very diverse timescales of the major processes involved.

The first three chapters address the basics: What are greenhouse gases and how do they lead to warming? How and why does the atmospheric warming vary in time and space (both as a function of latitude and height)? And finally, sea level rise. Beginning with greenhouse gases, the blue line in Figure 1.1a shows the iconic Mauna Loa CO_2 record collected since 1958, preceded by an ice-core-based reconstruction. CO_2 concentration has been at 280 ppm for over 10,000 yr, since the last ice age, and has therefore increased by about 50% so far, at an unprecedented speed. There is, of course, no doubt that CO_2 is increasing and that the increase is attributable to the anthropogenic burning of fossil

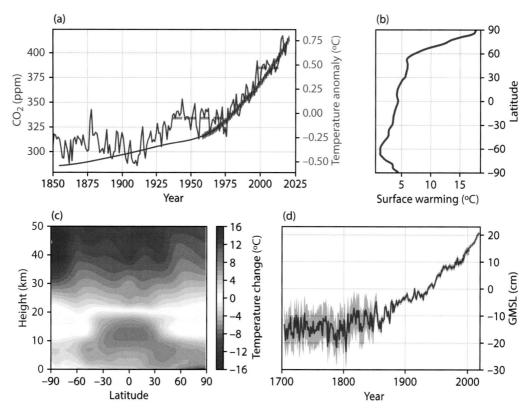

Figure 1.1: Greenhouse effect, warming, and sea level rise.
(a) Atmospheric CO_2 concentration (blue) and global mean surface temperature anomaly (in red, defined as the deviation from the mean over 1961–1990) since 1850. Also shown by gray dashed lines are two "hiatus" periods of *seemingly* reduced rates of warming. (b) The projected surface warming over the 21st century in an RCP8.5 scenario in a climate model, as a function of latitude, showing a very pronounced polar (mostly Arctic) amplification. (c) Projected atmospheric temperature changes over the 21st century as a function of altitude and latitude, showing a tropospheric warming and stratospheric cooling in the RCP8.5 scenario. (d) Estimated global-mean sea level anomaly since 1700 (blue line) and the estimated uncertainty (light-blue shading). In red and orange: the satellite record since 1993.

fuel. Once in the atmosphere, we will see that it will take thousands of years for the CO_2 to naturally decline after anthropogenic emissions are eliminated. Chapter 2 addresses the question of how greenhouse gases trap heat and lead to warming, both on a molecular level and by examining the Earth global energy balance. These are ideas that have been well understood for a while now. The warming due to greenhouse gas increase is explored in chapter 3. The red line in Figure 1.1a shows that the globally averaged surface temperature anomaly (defined as the deviation from a reference value, in this case the mean over 1961–1990) has warmed so far by over 1 °C. Future climate projections rely on estimates of future greenhouse gas concentrations, referred to as Representative Concentration Pathways (RCPs) and followed by a number indicating the expected enhancement in radiative heating. We discuss these scenarios in section 2.1 and note briefly now that RCP8.5 is what one might think of as a business-as-usual scenario in which CO_2 concentration increases to a very high value of over 1000 ppm by year 2100. While one hopes that such a high future greenhouse gas concentration is not a realistic scenario, it allows us to clearly understand climate change trends and mechanisms that may be more difficult to identify in less severe scenarios. Figure 1.1b shows that the projected surface warming under RCP8.5 is strongly amplified toward the poles (especially the Arctic), while Figure 1.1c shows that the stratosphere is projected under the same scenario to cool significantly above an altitude of about 20 km, while the troposphere warms. Polar amplification and stratospheric cooling are already observed today, and we will examine several mechanisms that are responsible for these signals. We will also see that even at the present level of CO_2, additional warming would have occurred if not for the cool, deep ocean, which takes hundreds of years to warm in response to the enhanced greenhouse forcing. While CO_2 has increased monotonically, the warming of the global mean surface temperature seems to have paused during 1940–1970 or so and during 1998–2013, as shown by the horizontal dashed lines in Figure 1.1a. We will show that such seeming "hiatus" periods in the increase in global mean surface temperature are an expected consequence of anthropogenic warming in the presence of natural climate variability.

One of the most consequential results of ocean warming and of the expected melting of land-based ice is sea level rise, as analyzed in chapter 4 and shown in Figure 1.1d. Global mean sea level has increased by some 30 cm over the past 150 yr, is currently increasing at about 3.5 mm/yr, and is projected to rise by up to a meter by 2100. Many processes are responsible for sea level rise, from the expansion of warming ocean water to land-based ice melting. Furthermore, sea level rise is expected to vary from region to region, and we will discuss the many mechanisms involved, from wind and atmospheric pressure changes, to the gravitational effect of melting ice over Greenland and Antarctica, and more. The timescales of the processes involved vary from a near-instantaneous

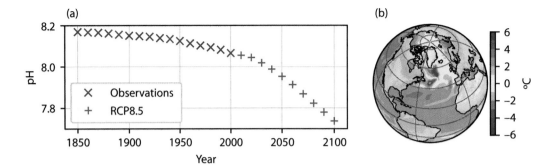

Figure 1.2: Ocean acidification and circulation.
(a) Observed mean surface ocean pH from 1850 to 2000 (blue), followed by projected pH to 2100 in the RCP8.5 scenario (red). (b) Projected sea surface temperature change over the 21st century in the RCP8.5 scenario, showing an overall warming, yet with a local cooling in the northern North Atlantic due to the projected collapse of the ocean overturning circulation.

response (e.g., to changes in atmospheric pressure or ocean currents) to hundreds of years (warming of the deep ocean, melting in Greenland), to many thousands (significant melting of East Antarctica), with uncertainty levels typically higher for processes with longer timescales.

Two more ocean-centered issues are next examined, beginning with ocean acidification in chapter 5 and then the possible collapse of the meridional overturning ocean circulation in chapter 6. Ocean acidification is referred to as "the other global warming problem." The absorption by the ocean of about a quarter of the anthropogenically emitted CO_2 (with another quarter absorbed by the land biosphere) has significantly reduced the warming experienced so far. Yet Figure 1.2a shows that as a result, ocean pH, the measure of seawater acidity, already decreased from 8.16 to 8.06, which implies a significant increase of 25%(!) in the concentration of H^+ ions in the ocean. We will examine the basic carbonate chemistry behind ocean acidification, how it may affect the deposition of calcium carbonate structures by oceanic organisms, and how atmospheric CO_2 can eventually decline once emissions are significantly reduced, on a timescale of thousands of years. While there is little uncertainty involved in assessing expected ocean pH for a given atmospheric CO_2 level, the response of ocean biology is complex and is still being studied.

Chapter 6 discusses how the oceanic meridional circulation, which carries heat poleward and contributes to the warmth of the high-latitude North Atlantic, may collapse in a global warming scenario over the next century or so. The circulation collapse is expected to contribute to a regionally reduced warming and even cooling in the northern North Atlantic, as shown in the model projection of Figure 1.2b, as well as to other disruptions to the current state of the ocean. We will analyze how a gradual CO_2 increase may lead to an abrupt ocean circulation response, explaining in the process the concept of climate *tipping points*.

Returning to the atmosphere in the next two chapters, we address two issues surrounded by a larger degree of uncertainty: clouds and hurricanes. In chapter 7 we study clouds, believed to be the main source of uncertainty in global warming projections and one of the main reasons that the uncertainty in global warming projections has not decreased over the past four decades. Unlike the discussion of climate change issues in other chapters, the focus of this chapter is not explaining an observed or projected change but rather making it clear why clouds are a source of such large uncertainty in our climate projections. Clouds have a most significant effect on climate due to both their reflectivity of sunlight, which has a cooling effect, and their trapping of heat emitted by the Earth surface, contributing to the greenhouse warming effect. Figures 1.3a,b show the projected change in clouds over the 21st century in the RCP8.5 scenario in two different climate models. The two models clearly calculate a very different cloud response, demonstrating the model disagreement and therefore the uncertainty in future projections of clouds. This disagreement in the simulation of cloud cover also leads to a very different warming projected by these two models, and we will explain why the representation of clouds in climate models involves such a large uncertainty. The subject of clouds allows us to also explore that of atmospheric convection, which comes up repeatedly in the discussion of many global warming–related issues. Following that, the response of hurricanes to global warming, both observed and projected for a future climate, is analyzed in chapter 8. Figure 1.3c shows the estimated number of Atlantic hurricanes over the past 140 yr. It is difficult to identify a trend in these data, and it turns out that there is currently no reliable and well-understood mechanism that can be used to project future changes to the *number* of storms. We discuss the formation mechanism of hurricanes, how it depends on the upper ocean temperature, and why the *magnitude* of hurricanes may be expected to increase in a warmer climate. We also analyze the observed record and examine the many uncertainty factors involved in the projection of future hurricane magnitudes.

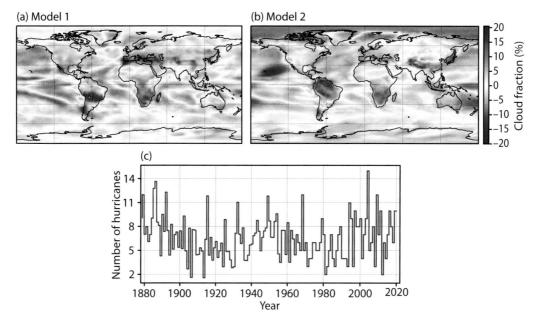

Figure 1.3: Clouds and hurricanes.
(a,b) Projected change in cloud cover due to greenhouse forcing, from a preindustrial state to 2100 in the RCP8.5 scenario, for two different climate models, demonstrating the large uncertainty in simulating clouds in climate models. (c) Estimated number of hurricanes over the North Atlantic as a function of year.

The following three chapters deal with the cryosphere: Arctic sea ice, ice sheets over Greenland and Antarctica, and mountain glaciers. Chapter 9 explains the processes and powerful feedbacks behind the dramatic and well-observed decline of summer Arctic sea ice over the past few decades (but not of sea ice near Antarctica, interestingly). This decline is seen in Figure 1.4a, and the same processes and feedbacks may lead to an even more dramatic decline over the next few decades. We also discuss ways of differentiating between a sea ice melt trend due to anthropogenic climate change and trends due to natural variability. Possible mechanisms and feedbacks that may lead to a significant reduction of the ice mass of the large ice sheets of Greenland and Antarctica are presented in chapter 10. Such a reduction can cause a further rise of sea level by many meters over a timescale of hundreds to thousands of years and involves a large degree of uncertainty. There is the (quite uncertain) potential for rapid changes as well, and we explain the mechanisms that may lead to such a tipping point behavior. This is followed by an analysis in chapter 11 of one of the more iconic

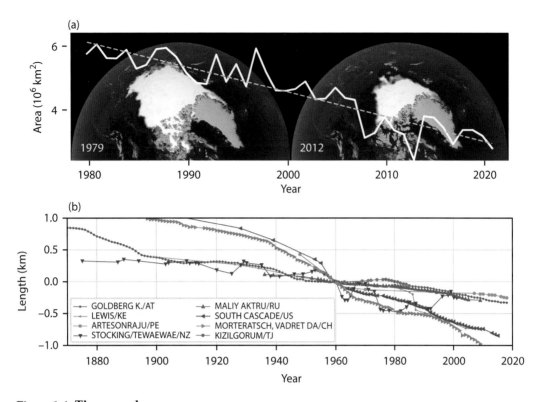

Figure 1.4: The cryosphere.
(a) The yearly minimum (September) Arctic sea ice area as a function of year over the satellite era, superimposed on NASA images of sea ice cover the first year of satellite data, 1979, and a year of a particularly small sea ice area, 2012. (b) Records of glacier length for a few mountain glaciers, relative to their length in 1960.

consequences of the already observed global warming: the retreat of mountain glaciers, as seen in Figure 1.4b. These changes are already occurring, and we will see that the retreat over the past three or so decades can be clearly attributed to anthropogenic global warming. We explain the processes behind this decline and the relevant dynamics of mountain glaciers that underlie their observed and projected decline.

We conclude with three chapters on possible consequences of climate warming, involving changes to droughts and precipitation, heat waves and forest fires. We review in chapter 12 the types and causes of prolonged droughts, as demonstrated in Figures 1.5a,b, showing how La Niña events in the equatorial east Pacific can lead to California droughts in a climate model. This is then used to consider why droughts might change in the future and why such a prediction

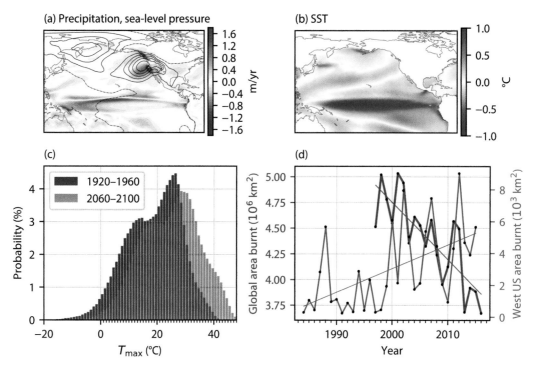

Figure 1.5: Droughts, heat waves, and forest fires.
(a,b) An analysis of droughts in a climate model. Colors in (a) show the precipitation anomaly during January drought years over California, with the blue area over California and off-coast there indicating lower precipitation than normal. The black contour lines show a high sea level atmospheric pressure anomaly that typically occurs above drought conditions (contour interval is 1 hPa; zero contour is dash-dot). (b) Sea surface temperature anomaly during these California drought years, showing a La Niña–like cold sea surface temperature (see box 8.1) over the equatorial Pacific and demonstrating how sea surface temperature anomalies drive remote drought conditions. (c) Heat waves in a warming climate. The probability of occurrence of maximum daily temperatures over the great plains in the United States at the beginning of the 20th century (blue) vs at the end of the 21st century in the RCP8.5 scenario (red). The shift to larger daily maximum temperatures is an example of projected changes to the characteristics of heat waves. (d) Forest fires. The red curve shows the increasing burnt area over the western United States over the past decades, and the blue curve shows a decrease in the global burnt area since the 1990s.

is still uncertain. We will see how the severity of past droughts can be reconstructed, helping to put current drought events into perspective, and we will model the processes that control soil moisture during droughts. We then analyze two test cases, the Sahel and Southwest United States, and explicitly demonstrate the uncertainty in future projections of droughts. Finally, we consider projections for precipitation changes from three perspectives. First, we discuss the projection that the Hadley meridional atmospheric circulation—of air rising near the equator and sinking at about 30° north and south—might expand poleward. The expansion has possible consequences to the location of desert bands that tend to be located under the subsiding part of the Hadley circulation. Second, we attempt to understand the projection that precipitation changes will follow a pattern of *wet getting wetter and dry getting drier* over large areas of the globe, as well as the limitations of this overall projected pattern, which does not seem robust over land areas. And finally, we examine projections for precipitation extremes and explain why there is a robust expectation for more precipitation to occur in heavy precipitation events in a warmer climate.

Heat waves are studied in chapter 13. These are weather events, and are therefore much shorter than droughts but share some of their physical mechanisms and characteristics. Figure 1.5c shows how the probability of occurrence of a high maximum daily temperature dramatically increases in model projections from the early 20th century (blue bars) to late in the 21st century (red). We demonstrate how the statistics of heat waves may change and what this can teach us about their dynamics in a warmer climate. We conclude with the subject of forest fires (chapter 14). Observations suggest a recent increase in fires over the western United States, for example, as seen in Figure 1.5d (red line), although global fire area has decreased over the past couple of decades (blue). We address factors affecting forest fires and different ways in which humans can affect fires via both climate- and non-climate-related influences. While we do understand qualitatively how fires depend on these many different factors, the issues involved are sufficiently complex that the only way to attempt to differentiate the effect of anthropogenic climate change from other anthropogenic effects and from natural climate variability is statistical analysis, resulting in very significant uncertainty. We discuss, as specific examples, fires in the western United States and Australia, as well as on a global scale, and attempt to identify many of the uncertainties involved.

As we move into the detailed analysis of these issues that arise in global warming science, we keep in mind the following questions: *What* has been observed or what is projected? *Why* do these changes occur, or why are they expected? What is the *timescale* in which these changes operate? What are the *uncertainty levels*, and *sources of uncertainty*?

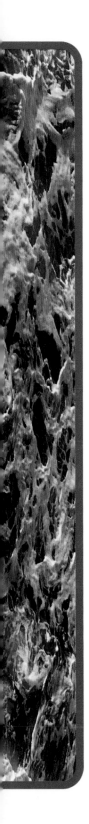

1.1 WORKSHOP

A Jupyter notebook with the workshop and corresponding data file are available; see https://press.princeton.edu/global-warming-science.

Go over and solve the first python notebook with an introduction to programming and a very brief review of some basic math concepts that are used later in the course.

GREENHOUSE

Camille Hankel and E.T.

Key concepts

- Energy balance, the greenhouse effect
- Emission height, atmospheric lapse rate, response to greenhouse gas increase
- Black body radiation
- Greenhouse gases, how they absorb radiation
  - Molecular vibrational and rotational modes
  - Energy levels, absorption lines, absorption windows
  - Pressure and Doppler broadening
  - Different greenhouses gases compared, greenhouse warming potential, CO_2-equivalent mixing ratio
- Water vapor feedback to increased CO_2

We begin with the basics: What are greenhouse gases, how do they absorb radiation, and why and how does this lead to warming? Observations clearly show that CO_2 has increased over the past century and a half due to the burning of fossil fuels, at a rate much faster than has occurred naturally in the past (Box 4.1), and some projections suggest that this rise may not stop or slow down dramatically in the near future. Figure 2.1 shows the observed CO_2 concentration time series (blue), the—hopefully—unrealistic high-end greenhouse gas concentration scenario known as the Representative Concentration Pathway (RCP) 8.5 used by the Intergovernmental Panel on Climate Change (IPCC), and some more moderate scenarios. The concentrations of other anthropogenic greenhouse gases have also increased, including methane (CH_4), nitrous oxide (N_2O), CFC12, and CFC11, which further increase the anthropogenic greenhouse effect.

The discussion starts with simple representations of the greenhouse effect, allowing us to understand how it leads to warming (section 2.1). We then consider how greenhouse gases interact with radiation and the role of the wavelength-dependence of the absorption of electromagnetic radiation (heat) by greenhouse gases (sections 2.2.1–2.2.4). We conclude with a comparison of different greenhouse gases (2.2.5) and with the water vapor

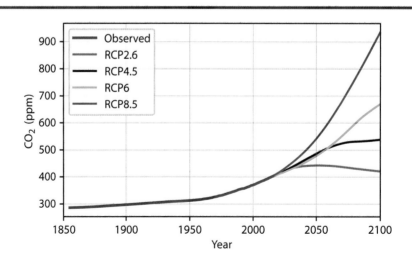

Figure 2.1: CO_2 timeseries.
Annually averaged CO_2 concentration, observed and projected according to different RCP scenarios.

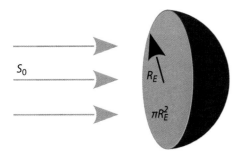

Figure 2.2: Solar forcing.
The total solar radiation distributed over the entire Earth surface is equal to the incoming solar radiation flux per unit area at the top of the atmosphere, S_0 (W/m^2), times the cross-section of the Earth, πR_E^2 (m^2), which is shown as the orange cross-section area, for a total of $S_0 \pi R_E^2$ W. This radiation is then distributed over the entire Earth area during one day due to the Earth rotation.

feedback that further amplifies the warming effect of anthropogenic greenhouse gases (2.2.6).

2.1 THE GREENHOUSE EFFECT
2.1.1 Earth's energy balance

We can estimate the globally averaged Earth surface temperature based on the balance of incoming radiation from the Sun with the outgoing heat escaping to outer space. Consider first what the averaged Earth surface temperature would have been in the absence of the radiative effects of the atmosphere. The incoming solar radiation at the top of the atmosphere is given by the solar constant, $S_0 = 1361$ W/m^2. The total solar radiation encountered by Earth is therefore the solar constant times the cross-section area of Earth (orange area in Figure 2.2), $S_0 \pi R_E^2$, where R_E is the Earth radius. Over a day, this radiation is distributed over the entire Earth surface area, $4\pi R_E^2$, so that the radiation per unit area, averaged over the Earth surface and over a day, is the ratio of the total radiation and the total area, or $S_0/4$. The fraction of the incoming solar radiation reflected (by ice, clouds, land, and so on), or the *albedo*, is denoted by $\alpha \approx 0.3$.

Both the Earth and the Sun emit radiation based on their corresponding surface temperatures. This emitted radiation is close to that of a *black body* at these temperatures. A black body is one that absorbs all incident radiation, and when it is in thermal equilibrium at a temperature T, it emits a total radiation per unit area of σT^4, where $\sigma = 5.670 \times 10^{-8}$ W m^{-2}K^{-4} is the Stefan-Boltzmann constant. Assume the outgoing radiation from Earth (escaping heat) to be given by the black body radiation formula, with T being the globally averaged surface temperature. The electromagnetic radiation from the Sun has wavelengths of about 0.25–2 micrometers (μm) and is referred to as *shortwave* (SW) radiation, which includes visible light that is characterized by wavelengths of 0.4–0.7 μm. The thermal radiation emitted by Earth is characterized by wavelengths of roughly 5–35 μm and is therefore referred to as *longwave* (LW) radiation.

Assuming the Earth to be in thermal equilibrium, the incoming shortwave radiation from the Sun that is not reflected must be equal to (i.e., balance) the outgoing longwave radiation (Figure 2.3a),

$$\frac{S_0}{4}(1 - \alpha) = \sigma T^4, \tag{2.1}$$

which gives

$$T = \left(\frac{(S_0/4)(1 - \alpha)}{\sigma}\right)^{1/4} = 255\,\mathrm{K} = -18\,^{\circ}\mathrm{C} \equiv T_0. \tag{2.2}$$

This is too cold, as at such a temperature the Earth surface would have been frozen, and the actual globally average surface temperature of the Earth is about 14 °C (287 K), so something is clearly missing. That missing factor is the greenhouse effect of the atmosphere.

2.1.2 The greenhouse effect: a two-layer model

We now add the greenhouse radiative effect of the atmosphere, whose temperature is denoted by T_a. To begin, we treat the atmosphere as a single layer and assume that it absorbs heat (longwave radiation) escaping from the surface and then re-emits it both up and down at a rate depending on the atmospheric temperature (Figure 2.3b). We write separate energy balance equations for the surface and for the atmosphere. The atmosphere is not a perfect black body and therefore only emits a fraction ϵ of the radiation of a black body with the same temperature. Similarly, it also absorbs a fraction ϵ of the LW radiation from the surface. This fraction is referred to as the LW *emissivity*, which is smaller than but close to one. The emissivity, also equal to the absorptivity (the part of incident radiation absorbed by a surface divided by that absorbed by a black body), is a function of the water vapor and CO_2 concentrations, among other things, and can be set for preindustrial climate to, say, 0.75. Thus the atmosphere emits LW

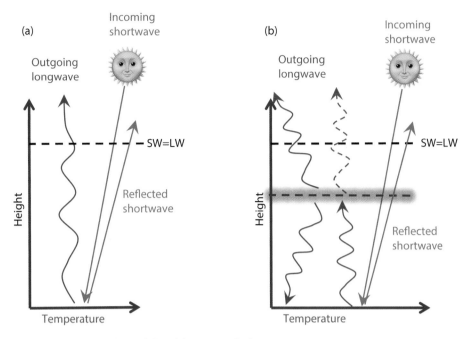

Figure 2.3: Two models of Earth's energy balance.
(a) A one-layer model ignoring the radiative effects of the atmosphere. (b) A two-layer model including the greenhouse effect of the atmosphere. Straight arrows indicate received and reflected SW radiation. Solid wiggly red lines denote LW radiation emitted by the surface and atmosphere; the dashed wiggly line denotes the part of the surface-emitted LW radiation not absorbed by the atmosphere.

radiation at a rate of $\epsilon \sigma T_a^4$ W/m^2 both upward and downward. As the surface (whose emissivity is closer to one) radiates σT^4 W/m^2, the atmosphere absorbs only a fraction ϵ of this radiation, with the rest continuing to outer space (wiggly dashed upward arrow in Figure 2.3b). Given these assumptions, the energy balances for the surface and atmosphere may be written as

$$\frac{S_0}{4}(1 - \alpha) + \epsilon \sigma T_a^4 = \sigma T^4$$

$$\epsilon \sigma T^4 = 2\epsilon \sigma T_a^4. \tag{2.3}$$

The first equation is the energy balance for the surface, showing the input from the Sun and from the atmosphere on the LHS and the output on the RHS. The second equation represents the energy balance for the atmosphere, with input

LW radiation from the surface on the LHS and output due to the LW radiation from the atmosphere on the RHS. The factor 2 on the RHS represents the upward and downward LW emission from the upper and lower surfaces of the assumed single-layer atmosphere. Note that the atmosphere is assumed transparent to the shortwave radiation from the Sun, a reasonable assumption to first order, so that the solar radiation warms the surface, and the radiation from the surface then affects the atmosphere. Substitute the second equation in the first,

$$\frac{S_0}{4}(1-\alpha) + \frac{1}{2}\epsilon\sigma T^4 = \sigma T^4,$$

so that

$$T = \left(\frac{(S_0/4)(1-\alpha)}{\sigma(1-\epsilon/2)}\right)^{1/4} = T_0(1-\epsilon/2)^{-1/4} = 284\,\mathrm{K} = 13\,^\circ\mathrm{C}, \quad (2.4)$$

which is reasonably close to the observed global mean surface temperature. Had we assumed the atmosphere to be a perfect black body, the corresponding solution for $\epsilon = 1$ is $T = T_0 2^{1/4} = 303\,\mathrm{K} = 30\,^\circ\mathrm{C}$, which is too warm. The difference between the energy balances with and without an atmospheric layer taken into account (panel a vs panel b in Figure 2.3, and eqn 2.2 vs eqn 2.4) demonstrates the greenhouse effect of the atmosphere. It should be noted that water vapor is the main atmospheric greenhouse gas that contributes to the LW emissivity/absorptivity of the atmosphere and accounts for most of the greenhouse effect, although with a critical contribution from CO_2.

Finally, getting to the anthropogenic greenhouse effect, we note that an increased concentration of greenhouse gases increases the LW absorptivity/emissivity of the atmosphere, ϵ, and the solution (eqn 2.4) shows that the surface temperature will increase accordingly. It turns out, though, that while the explanation of the natural greenhouse effect of the atmosphere using this two-layer model is helpful, this is too much of an oversimplification when it comes to the atmospheric response to the anthropogenic increase in greenhouse gas concentrations. This picture is accordingly refined in the next subsection.

It is worth noting, perhaps, that an actual greenhouse, such as that shown at the beginning of this chapter, operates differently from the above atmospheric greenhouse effect. The glass allows sunlight in, and while it prevents much of the infrared (LW) radiation from the surfaces inside the greenhouse from escaping due to its optical properties, this is not the main mechanism by which it keeps the air in the greenhouse warm. The glass temperature is not significantly colder than that of the air inside the greenhouse, and thus the LW radiation it emits is not much lower than that of the air in the greenhouse. Similarly, while glass is a relatively good insulator, the typical thickness of about 3 mm used for constructing greenhouses cannot significantly reduce conductive (diffusive) cooling through the glass. Instead, the glass keeps the greenhouse warm by preventing air exchanges with the surroundings: surfaces under the glass absorb solar radiation and warm, also warming the interior air that is in contact with these surfaces. The glass, then, prevents the warm air from rising and being replaced by (or mixed with) cooler air, as would have happened outside the greenhouse, keeping the interior warm. In the atmosphere, greenhouses gases trap longwave radiation rather than restricting the movement of warm air.

2.1.3 The emission height and lapse rate

The absorption by the atmosphere of longwave radiation emitted by the surface depends on the wavelength of the radiation, as further discussed in section 2.2. This differential absorption depends on the concentration of different greenhouse gases in the atmosphere, each of which absorbs at different wavelengths. It turns out that the atmosphere absorbs nearly all photons of longwave radiation emitted from the surface in the wavelength range corresponding to CO_2 absorption. Why is it, then, that increasing CO_2 concentration is still expected to lead to further warming? One of the main reasons is that the temperature declines with height in the lower few kilometers of the atmosphere, rather than the atmosphere being a single layer with a single temperature, as assumed in section 2.1.2.

The atmospheric temperature decreases linearly with height in the troposphere at a *lapse rate* of approximately 5–9.8 °C per km (this will be derived in section 7.2) and then increases in the stratosphere (Figure 3.8a) due to absorption of SW solar radiation by the ozone layer there. Most of the radiation emitted from Earth's surface is absorbed by the air above it, which re-emits the radiation both downward and upward to the air above, and so on. At some level in the atmosphere, the overlying air is thin enough that most of the longwave radiation emitted from that level is able to make it to outer space without being absorbed again. This is the *emission height*, and the air temperature at this height determines the flux of outgoing longwave radiation from the atmosphere. Increasing the concentration of greenhouse gases moves the emission height farther up, as some of the radiation from the previous emission height is trapped by the newly added greenhouse gas molecules above this height. Doubling the CO_2 concentration, for example, raises the emission height by about 150 m. Given that there is only weak LW absorption above it, the temperature of the emission height at equilibrium, once the climate system adjusts to the changed greenhouse gas concentration, is determined by the balance between incoming shortwave radiation and outgoing longwave radiation from this height.

Figure 2.4 shows a schematic of the temperature profile and emission height (a) before an increase of the CO_2 concentration, (b) immediately after an assumed abrupt increase, and (c) after the atmospheric temperature adjusted. Before the increase in greenhouse gas concentration (a), the atmosphere emits LW radiation from the emission height to space at a temperature for which the outgoing longwave radiation (OLR) is equal to the incoming shortwave radiation (as in eqn 2.1). After the CO_2 increase, the emission height moves upward, and because of the lapse rate, the temperature that radiates to outer space is now lower, and therefore so is the OLR. Thus, the incoming shortwave radiation is greater than the OLR, causing an energy imbalance (b) that leads to a warming of the atmosphere. This warming causes the temperature profile to adjust (the solid green line moves to the right in Figure 2.4c) until the temperature at the new emission height is high enough to put the OLR in balance with incoming shortwave radiation again. This increase in the emission height determines the response of the energy balance at the top of the atmosphere to higher CO_2, leading to atmospheric warming. The surface is also warmed by a smaller, separate effect due to the increase in downward LW radiation at the bottom of the atmosphere, even before taking into account atmospheric warming: the increased atmospheric LW emissivity due to a higher CO_2 concentration implies that more of the downward LW at the surface arrives from lower in the atmosphere, where the temperature is higher; the higher emitting temperature means more LW radiation arriving at the surface and therefore a surface warming.

The key takeaway is that an increase in greenhouse gas concentration leads to an increase in the emission height and therefore to a warming at the surface. The

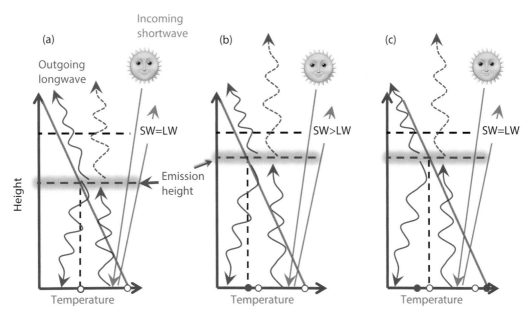

Figure 2.4: Role of the emission height and lapse rate in the greenhouse effect.
Emission height (dashed blue horizontal lines) and vertical temperature profiles (solid green) are shown (a) before a CO_2 increase, (b) immediately after a CO_2 increase but before the atmospheric temperature adjusted, and (c) after the atmospheric temperature adjusts. Red dots at the bottom of each panel denote the surface temperature. Blue dots denote the emission temperature. Before the warming and after the CO_2 increase (panel b), the emission temperature decreases (moving from the empty to the filled blue circle). As a result, the outgoing LW radiation decreases and is smaller than the incoming SW radiation. After the warming (panel c), the new emission temperature at the raised emission level is equal to the original emission temperature at the lower original emission level, denoted by the empty blue circle, thus balancing the incoming SW radiation again.

important factors leading to this warming are that the emission height depends on the CO_2 concentration and that the atmospheric temperature decreases with altitude within the troposphere. The lapse rate is further discussed and derived in section 7.2.

2.2 GREENHOUSE GASES

2.2.1 Wavelength-dependent black-body radiation

While we have been treating the outgoing radiation so far as a single entity, it is made of different wavelengths of electromagnetic radiation. The Earth emits from the surface approximately like a black body with a temperature of 288 K or so (blue curve in Figure 2.5a), while the Sun radiates as a black body of a much higher temperature (red curve), where both curves follow Planck's Law,

$$B(\lambda, T) = \frac{2hc^2}{\lambda^5} \frac{1}{e^{\frac{hc}{\lambda k_B T}} - 1}. \tag{2.5}$$

Here, $B(\lambda, T)$ is the spectral radiance emitted by the black body at a wavelength λ (m), at thermal equilibrium at temperature T (K). The radiance is the radiation in watts (W), per unit area (m^2) of the emitting surface, per unit solid angle (steradian), per unit wavelength (m): together, W·m^{-2}·ster^{-1}·m^{-1}. Also, $k_B = 1.38064852 \times 10^{-23}$ m^2 kg s^{-2} K^{-1} is the Boltzmann constant, $h = 6.62607004 \times 10^{-34}$ m^{2}kg s^{-1} the Planck constant, and $c = 3 \times 10^8$ m s^{-1} the speed of light. The integral of the spectral radiance over all wavelengths and angles yields the total black-body radiation, σT^4 (W m^{-2}), used previously, and σ can be expressed in terms of the constants appearing in equation (2.5).

2.2.2 Energy levels and absorption

A given molecule may be at different energy levels, depending on the energy levels occupied by its electrons and on which rotation and vibration modes of the molecular bonds are excited. A photon of frequency ν and wavelength $\lambda = c/\nu$ has an energy $h\nu$. For specific frequencies, the photon may be able to deliver the exact amount of energy needed to cause the molecule to transition between its energy level and another level and thus be absorbed by the molecule. Moving electrons to higher orbitals typically requires very energetic photons in the ultraviolet part of the SW range. Exciting vibrations and rotations of greenhouse gas molecules in the atmosphere requires photons in the LW range, allowing these molecules to absorb LW radiation. Photon frequencies or wavelengths corresponding to the absorption energies of greenhouse gases are referred to as absorption lines. Figure 2.5b shows the estimated top of the atmosphere outgoing LW radiation; one can clearly identify wavelength bands where the OLR is significantly below the 287 K black body curve representing emission from Earth's surface. These bands correspond to wavelengths in which different greenhouse gases absorb and re-emit outgoing radiation, eventually

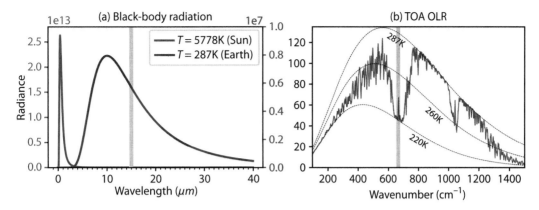

Figure 2.5: Black-body radiation.
(a) Planck's black-body spectral radiance $B(\lambda, T)$ for the emission temperatures of the Earth (blue) and Sun (red), as a function of wavelength, in $W \cdot m^{-2} \cdot ster^{-1} \cdot m^{-1}$. The Sun radiates much more strongly, of course, and its peak emission is at a shorter wavelength (hence "shortwave" radiation), whereas Earth's peak emission comes at longer wavelengths ("longwave radiation"). (b) The blue curve shows an estimated outgoing longwave radiation at the top of the atmosphere as a function of wavenumber, with black-body radiation curves at different temperatures shown by dashed lines: $mW\, m^{-2}\, ster^{-1}\, (cm^{-1})^{-1}$. The deviations from the 287 K black body radiation curve indicate absorption bands due to CO_2, CH_4, H_2O, and O_3; the central CO_2 absorption line is shown on both panels as a vertical gray bar. The two panels use different units for the x-axis that are often used to display such spectra, wavelength (λ, μm) on the left and wavenumber ($1/\lambda$, cm^{-1}) on the right.

re-emitting it from a higher altitude and thus at a colder temperature to outer space, and therefore reducing the OLR at these wavelengths. The absorption from 400 to 600 cm^{-1} and beyond 1200 cm^{-1} is due to water vapor, which is the dominant greenhouse gas, responsible for approximately 50% of the Earth longwave greenhouse effect. The absorption from 1000 to 1100 cm^{-1} is due to ozone, and the minimum in OLR around 600–800 cm^{-1} marked by the gray bar (also shown by the gray bar at 15 μm in panel a) is due to absorption by CO_2.

2.2.3 Broadening

Molecules can absorb photons with energy in a broader range around the precise energies corresponding to transitions between pairs of molecular energy levels due to two main effects. The first is *pressure broadening*, which may occur when two molecules collide while one of them is absorbing a photon. In that case, as one molecule absorbs the photon, some excess energy can be transferred to/from the colliding molecule, allowing the absorbing molecule to retain exactly the energy needed for an energy-level transition. As a result, the absorbed photon does not need to be at exactly the energy corresponding to the energy-level transition, showing up as a broadening of the lines of absorbed photon frequencies. This broadening is enhanced at higher pressures, as the frequency of molecular collisions is enhanced then—hence the name of the effect.

The second (and significantly weaker effect in the Earth atmosphere) is Doppler broadening: the thermal motion of molecules leads to a shift in the frequency of emitted/absorbed photons as seen by the moving molecule due to the Doppler effect. A photon with a slightly different frequency than that needed to induce a transition in a resting molecule can therefore seem by the moving molecule to have exactly the right frequency, allowing the moving molecule to absorb it.

Given these two effects, absorption is maximal at the absorption lines, at the frequencies/wavelengths corresponding to molecular energy transitions, and it gradually decays away from these lines. Even when the atmosphere is saturated with respect to absorption at the CO_2 absorption lines, increasing CO_2 concentration leads to stronger absorption at wavelengths near the absorption lines that are not saturated. The increased LW absorption near these lines again explains the greenhouse effect due to increasing CO_2 concentration, in addition to the increase in the emission height discussed in section 2.1.3.

2.2.4 Radiative forcing, logarithmic dependence on CO_2

Consider radiation propagating through a pipe filled with a radiation-absorbing gas (say CO_2), and let $I(x)$ be the intensity (W/m^2) of the *transmitted* radiation as a function of distance x. The transmitted radiation intensity is expected to decline along the pipe due to the absorption, leading to a negative rate of change, dI/dx, along the length of the pipe. The absorption is typically proportional to the radiation intensity at any given location, and therefore the equation for the transmitted radiation, $I(x)$, may be written as $dI/dx = -\mu I$, where μ

depends on the absorbing medium. The solution for the radiation intensity along the pipe is therefore $I(x) = I_0 e^{-\mu x}$, so that the transmitted radiation decreases exponentially with distance and with the amount of CO_2 encountered by the radiation.

The *radiative forcing* due to a given atmospheric CO_2 increase is the radiation (in W/m^2) that is absorbed by the added gas and is prevented from being transmitted to outer space, over all wavelengths, assuming the tropospheric temperature, moisture, and clouds are kept at their values before the CO_2 increase. Radiative forcing is calculated using radiation models rather than being observed directly, yet the uncertainty involved is small relative to many other factors affecting the climate response to anthropogenic greenhouse forcing.

Exact calculations show that the radiative forcing of CO_2 depends *logarithmically* rather than exponentially on CO_2 concentration. That the response is not exponential is a result of the fact that, as discussed above, the atmosphere is saturated with respect to CO_2 absorption—that is, LW photons emitted from the surface at the wavelengths of the CO_2 absorption lines are already absorbed by the atmosphere at preindustrial CO_2 concentrations. The shape of the absorption bands as a function of wavelength, and as a function of the distance from the absorption lines discussed above, leads to the logarithmic dependence of radiative forcing on the CO_2 concentration. We show in section 3.2 that the warming and radiative forcing are approximately linearly related. This implies that the temperature response also depends logarithmically on the CO_2 concentration. We therefore write, schematically, $T = T_0 + A \log_2(CO_2/280)$, where the use of log base 2 implies that A (in Kelvin) is the temperature increase if CO_2 concentration doubles from its preindustrial value of 280 ppm. This dependence means that the warmings due to one and two doublings of the CO_2 concentration are

$$T_{\times 2} - T_{\times 1} = A \log_2(2CO_2/280) - A \log_2(CO_2/280) = A \log_2 2 = A,$$

$$T_{\times 4} - T_{\times 2} = A \log_2(4CO_2/280) - A \log_2(2CO_2/280) = A \log_2 2 = A.$$

That is, each doubling of CO_2 leads to the same increase in temperature!

2.2.5 Other greenhouse gases, global warming potential

Human activities led to the increase in concentration of different greenhouse gases that can absorb in the longwave range, including carbon dioxide (CO_2), methane (CH_4), nitrous oxide (N_2O), CFC12, CFC11, and more. By 2020, the radiative forcing due to CO_2 relative to 1750 has been estimated at about 2 W/m^2, and when taking into account the other anthropogenic greenhouse gases mentioned above, the total radiative forcing is about 3 W/m^2. For perspective, the RCP8.5 scenario, representing a business-as-usual worst-case scenario, corresponds to a radiative forcing increase of 8.5 W/m^2 by year 2100. One way to present the effect of other greenhouse gases is via their CO_2-equivalent mixing ratio, which is the CO_2 concentration that would lead to the same radiative forcing as a given concentration of another given greenhouse gas. While the concentration of CO_2 just passed 400 ppm by 2020, the combined CO_2-equivalent mixing ratio due to all greenhouse gases reached 500 ppm.

Some of these added anthropogenic gases have a larger radiative forcing than CO_2 per kilogram of gas, yet their lifetime in the atmosphere may be shorter. For example, an added unit weight of methane (CH_4) leads to a significantly stronger increase in LW absorption than the same weight of added CO_2. But methane has a shorter lifetime in the atmosphere (about a decade; it eventually reacts with the OH radical to form CO_2 and water) than CO_2 (thousands of years; see section 5.3.3). It is useful, therefore, to be able to compare the *long-term* radiative effects of different gases, not reflected by the CO_2-equivalent mixing ratio; this is done using the concept of global warming potential (GWP). The GWP of a greenhouse gas is its forcing effect relative to that of CO_2, taking into account both its lifetime and its strength as an absorber, over a specified period referred to as a *time horizon* (TH). The GWP of a gas x is

$$GWP(x) = \frac{\int_0^{TH} a_x \cdot [x(t)] \, dt}{\int_0^{TH} a_r \cdot [r(t)] \, dt},$$

where $x(t)$ is the time-dependent decay in abundance of the gas and $r(t)$ that of CO_2; a_x is the radiative forcing of x per 1 added kilogram of x (W/m^2/kg), and a_r is that of CO_2.

Atmospheric water vapor plays a dominant role in the response to greenhouse gas increase, as well as in the many consequences of warming, from anticipated changes to droughts, forest fires, hurricanes, and extreme precipitation events. The saturation water vapor pressure is the pressure of water vapor in equilibrium with water at a given temperature. The relative humidity is defined as the ratio of the water vapor pressure and saturation water vapor pressure. To a first approximation, the relative humidity may often be assumed not to change in response to warming, which means that the water vapor response is determined by the behavior of the saturation moisture as a function of temperature, given by the Clausius-Clapeyron relation.

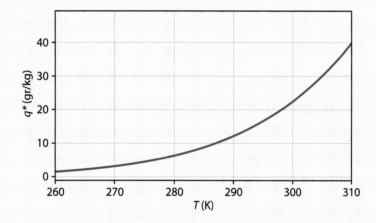

The saturation specific humidity, denoted $q^*(T, p)$, is the mass (kg) of moisture in 1 kg of moist air, at saturation, at a temperature T in degrees Celsius and air pressure p in hPa, and may be calculated as follows. Let $R_v = 461$ J K^{-1} kg^{-1} be the specific gas constant for water vapor and $R_d = 287$ J K^{-1} kg^{-1} the specific gas constant for dry air. The saturation water vapor pressure ($e^*(T)$, in hPa) is then given by the approximate Clausius-Clapeyron relation $e^*(T) = 6.112 \exp (17.67T/(T + 243.5))$. The saturation mixing ratio (kg water vapor per kg dry air) is given by $r^* = (R_d/R_v)e^*/(p - e^*)$, while the saturation specific humidity (kg water vapor per kg moist air) is, finally, $q^* = r^*/(1 + r^*)$. Note that $q^*(T, p)$ is exponential in the temperature as shown in the figure, a strong dependence that is behind the important role of atmospheric moisture in climate change.

The above Clausius-Clapeyron relation is an approximate solution to the Clausius-Clapeyron equation,

$$\frac{1}{e^*(T)} \frac{de^*(T)}{dT} = \frac{L}{RT^2} \equiv \alpha_{cc}(T),$$

where $L \approx 2250 \, \text{J K}^{-1} \text{kg}^{-1}$ is the (temperature-dependent) latent heat of vaporization, and α_{cc} is the expected relative change to the saturation moisture per degree Kelvin warming. This equation is also useful on its own as it shows that a $\Delta T = 1 \, \text{K}$ warming leads to a 7% increase in moisture: $\Delta e^*/e^* = \alpha_{cc}(T) \Delta T \approx 0.07$. Therefore, a mid-range estimated surface warming of 3 K in response to a doubled CO_2 concentration implies an over 20% increase in water vapor pressure or in specific humidity. The consequences of such a dramatic increase are reviewed throughout this book where the Clausius-Clapeyron relation is repeatedly used. Finally, we note that for a typical surface pressure of $p_s = 1000 \, \text{hPa}$, the saturation specific humidity may be further approximated as a function of the temperature T in Kelvin, as $q^*(T, 1000 \, \text{hPa}) = 1.58 \times 10^6 \exp(-5415/T)$ (kg moisture per kg moist air).

2.2.6 The water vapor feedback

Greenhouse gases such as CO_2 and methane also have an indirect radiative effect. A warmer atmosphere due to an increased greenhouse gas concentration allows for more atmospheric water vapor content in accordance with the Clausius-Clapeyron relation (Box 2.1). Since water vapor is also a potent greenhouse gas, this causes further warming, which causes the moisture content to further increase and the temperature to further increase. This amounts to a positive feedback: the initial warming leads to more water vapor, which further amplifies the warming. Model calculations and physical arguments show that the relative humidity may be assumed to remain roughly unchanged as climate warms, allowing us to calculate the approximate expected increase in specific humidity directly from the Clausius-Clapeyron relation. Estimates are that a surface warming of $1\,^{\circ}$C (say, due to the increase of CO_2) leads to increased atmospheric humidity that accounts for an extra radiative forcing of 2 W/m^2. For perspective, CO_2 doubling induces a radiative forcing of about 4 W/m^2. Put differently, the water vapor feedback roughly doubles the climate sensitivity, defined as the equilibrium surface warming response to the doubling of CO_2 (more in section 3.1).

Water vapor changes are a response (feedback) to an increase in greenhouse gas concentration rather than a direct forcing, because water vapor concentration responds to temperature changes on a very short timescale via surface evaporation and precipitation. Any climate warming involves a strong negative feedback due to the outgoing LW radiation (cooling) increasing as T^4, and thus leads to stronger radiative cooling in response to the climate warming. The positive water vapor feedback in the Earth atmosphere only weakens this negative feedback by weakening the increase in LW radiation as the temperature increases, such that the net feedback is still negative.

2.3 WORKSHOP

A Jupyter notebook with the workshop and corresponding data file are available; see https://press.princeton.edu/global-warming-science.

1. **Observed and projected increase in greenhouse gases:** Plot the CO_2 time series since 1850; superimpose the projected CO_2 from the RCP8.5 scenario until year 2100.

2. **Energy balance model:**

   (a) Calculate the downwelling LW and absorbed SW at the surface in the two-layer model.

   (b) Modify the two-layer energy balance model to take into account the absorption by the atmosphere of a fraction $\Delta = 0.15$ of the downward shortwave radiation, using an albedo of 0.25 and an atmospheric emissivity of 0.8. Calculate the resulting temperatures.

   (c) Consider an abrupt CO_2 doubling corresponding to the transition from panel a of Figure 2.4 to panel b. Calculate the increase in the emission height corresponding to an increase in radiative forcing by 4 W m^{-2} assuming a lapse rate of 6.5 K/km.

3. **Radiative forcing:**

   (a) Explain the upper and lower boundaries of the trapezoidal gap in the OLR curve around 600–800 cm^{-1} (Figure 2.5b) due to CO_2 absorption, in terms of the emission height.

   (b) Estimate the radiative forcing of 280 ppm of CO_2 by fitting a trapezoid to the gap in the OLR curve and calculating its area.

   (c) Plot the OLR in a standard atmosphere in the presence of 280 and 560 ppm CO_2, ignoring all other greenhouse gases.

   (d) Calculate the radiative forcing due to a doubling of the CO_2 concentration from the preindustrial value of 280 ppm, by integrating over the difference of the two OLR curves.

4. **Logarithmic dependence:** If a doubling of CO_2 from 280 ppm to 560 leads to, say, 3 °C warming, what warming do you expect at an equivalent CO_2 mixing ratio 500 ppm? Discuss the fact that warming so far is about 1 °C.

5. **Global warming potential:** Suppose the GWP of methane for a time horizon of 100 yr is 34. How much more efficient is a kilogram of methane at absorbing LW radiation than a kilogram of CO_2? Assume methane concentration decays exponentially over 12.4 yr

(so the decay function is $e^{-t/12.4}$) while that of CO_2 decays over 200 yr (e.g., due to dissolution in the ocean; note that while such a timescale is often used, the actual expected residence time of anthropogenic CO_2 in the climate system is thousands of years; see section 5.3.3). Discuss the difference between GWP and absorption efficiency. What is the methane GWP for a time horizon of 500 yr? Explain.

6. *Optional extra credit:* **Water vapor feedback to increased CO_2:** Modify the energy balance model equations (2.3) to include the effects of increasing water vapor, and calculate the warming as a function of CO_2 with and without this feedback as follows. Assume the atmospheric emissivity depends on CO_2 and specific humidity $q(T_a)$ as $\epsilon(CO_2, q(T_a)) = \epsilon_0 + A\log_2(CO_2/280) + B\log_2(q(T_a)/q(T_{a0}))$, where $\epsilon_0 = 0.75$ and T_{a0} is the atmospheric temperature for $CO_2 = 280$ ppm, and that the atmospheric relative humidity is fixed at 80%. Find values for A and B such that the warming due to CO_2 doubling alone is $2\,°C$, and due to both feedbacks is $4\,°C$. Plot $T_a(CO_2)$ for CO_2 in the range of 200 to 800 ppm with and without the water vapor feedback.

7. **Guiding questions to be addressed in your report:**

   (a) What does it mean that the atmosphere is already saturated with respect to LW absorption at the CO_2 absorption bands? Why is temperature still increasing in response to CO_2 increase?

   (b) What is the radiative forcing of CO_2, and how is it calculated?

   (c) Should we worry about CO_2 increase given that H_2O is a much more powerful greenhouse gas?

   (d) What are the sources of uncertainty in a calculation of radiative response to CO_2 increase by year 2100?

   (e) How is the temperature expected to increase as the CO_2 doubles once? Twice?

   (f) How do we quantify the effects of other greenhouse gases, and how would you use that to set a policy regarding the use of coal versus natural gas?

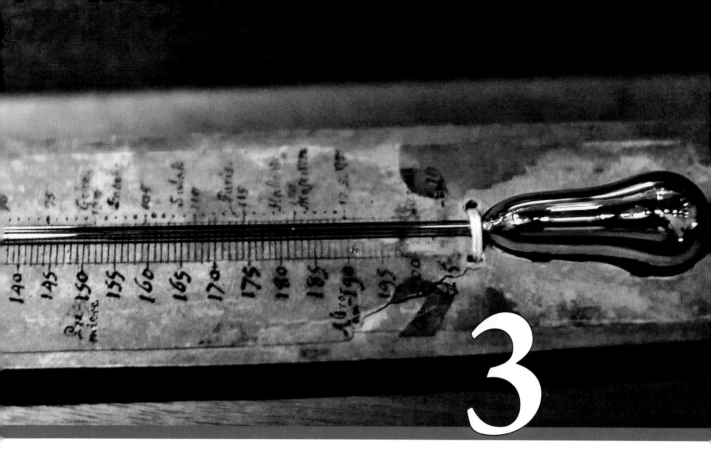

TEMPERATURE

Key concepts

- Equilibrium climate sensitivity
- Transient climate sensitivity and the role of the ocean
- Polar amplification
- Natural variability and "hiatus" periods
- Stratospheric cooling

The warming of surface temperatures over the past decades, and the expected further warming are, of course, at the heart of anthropogenic climate change and are the reason for many of the other issues related to global warming that affect us directly and indirectly. Figure 3.1 shows a time series of the observed globally averaged surface temperature anomaly (relative to the mean of

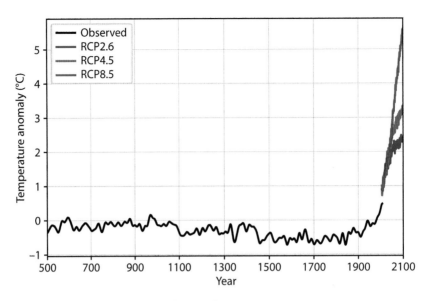

Figure 3.1: Observed and projected warming.
Observed globally averaged surface temperature anomaly (relative to the mean of 1961–1990) since year 500 (black) plus projections from three RCP scenarios in a climate model.

1961–1990) since the year 500, and the projected temperature time series for three different warming scenarios in a climate model. The observed record has been referred to as the *hockey-stick curve* due to its shape, showing a decline over most of the record and an increase in the past century or so. Climate models that are run under a CO_2 doubling scenario (from a preindustrial value of 280 to 560 ppm) predict a globally averaged warming of 2–4 °C as a result of the greenhouse forcing and all modeled feedbacks. By 2020, CO_2 already increased from 280 to nearly 420 ppm (and the CO_2-equivalent mixing ratio, which includes the effects of other anthropogenic greenhouse gases, has already reached 500 ppm; see section 2.2.5). We have seen that the dependence of radiative forcing on CO_2 is logarithmic (section 2.2.4), which seems to suggest that we should have seen most of the warming that is expected for CO_2 doubling. One therefore wonders if the warming that has so far been observed, of about 1.1 °C, is consistent with the climate warming projected for a doubling of CO_2. It turns out that the oceans play a dominant role in delaying the warming response to greenhouse gas increase; we will explore this issue in section 3.1.

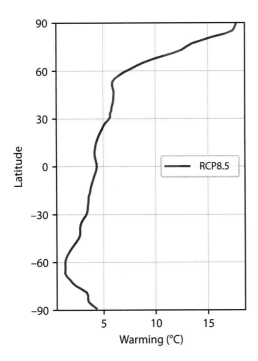

Figure 3.2: Polar amplification.
Warming over the 21st century as a function of latitude, based on the RCP8.5 scenario (Figure 2.1) in a climate model.

Figure 3.2 shows that the projected warming according to the RCP8.5 scenario is much larger in the Arctic than in mid-latitudes and that high latitudes in general are expected to warm more than low latitudes. This is referred to as *polar amplification*, and three mechanisms for this robust observation are discussed in section 3.2.

The warming over the past century is characterized by a couple of periods of weak warming or even weakly cooling trends, sometimes referred to as "*hiatuses.*" While the specific mechanisms for these periods of weaker warming trends are still hotly debated, they are an unavoidable result of combining natural climate variability with a warming trend, as discussed in section 3.3. Finally, a robust signature of CO_2-induced climate change is a cooling of the stratosphere (the atmospheric layer from 10 to 50 km above the surface), while the troposphere, the lower 10 km of the atmosphere, is warming. In section 3.4 we extend the two-layer atmospheric energy balance discussed previously to a three-layer atmospheric energy balance model to explain this.

3.1 CLIMATE SENSITIVITY AND THE ROLE OF THE OCEAN

Suppose the expected long-term surface warming after the climate system equilibrated with a doubling of CO_2 (or more precisely, a doubling of the CO_2-equivalent mixing ratio; see section 2.2.5), with all feedbacks taken into account, is $3\,°C$, roughly in the middle of the IPCC range. Given that we expect warming to be logarithmic in CO_2 concentration, we write the expected surface warming as a function of the CO_2 concentration as $\Delta T = 3 \log_2(CO_2/280)$. For a recently achieved equivalent CO_2 mixing ratio of 500 ppm, one therefore expects a warming of $\Delta T = 3 \log_2(500/280) = 2.5\,°C$, more than twice as much as has been observed. This section discusses and explains this seeming discrepancy in terms of the concept of *transient* climate sensitivity, after discussing first the related concept of *equilibrium* climate sensitivity.

3.1.1 Equilibrium climate sensitivity

Consider first how we may estimate the eventual warming of the climate system, after it had equilibrated in response to a doubling of the CO_2 concentration, using present-day observations. The increase in greenhouse gas concentrations implies an additional downward heat flux from the atmosphere toward the surface, which leads to a warming of the lower atmosphere and upper ocean (say upper 50 m), by an amount ΔT. We term this additional heat flux *radiative forcing* (measured in W/m^2, defined in section 2.2.4), denote it ΔF, and refer to the warming of the surface atmosphere and upper ocean as surface warming. The heat capacity of an object is the amount of heat (J) required to raise its temperature by one K and its units are therefore J/K. Note that the heat capacity of the atmosphere is of the same order of magnitude as the heat capacity of the upper 10 m of the ocean. The ocean, which is able to (slowly) mix heat down to great depths of up to 4000 m, therefore dominates heat budget changes due to an increase in greenhouse gas concentration. It is estimated that the oceans so far have absorbed about 90% of the excess heat due to increased greenhouse gas concentrations. This excess heat is defined to be the radiative forcing integrated over time. As we will see, this does not mean that the ocean *reduces* the eventual warming, but it can certainly *delay* it. While the land covering a third of the Earth surface is massive, of course, heat penetrates into soil or rocks only by very slow diffusion. Seasonal temperature variations can therefore penetrate about a meter into the land surface, while diurnal variations, having a much shorter timescale, penetrate only less than 10 cm. As a result, only the uppermost part of the land surface contributes to the effective heat capacity of the Earth surface, and its contribution to the storage of excess heat due to radiative forcing is significantly less than that of the ocean.

Climate models estimate that the increase in outgoing top-of-the-atmosphere (TOA) radiative cooling due to a surface warming ΔT is, to a good approximation, linearly proportional to this warming and may therefore be written as $\lambda_{LW} \Delta T$. This linearity is further explained in equation (3.5). The parameter λ_{LW} ($W\,m^{-2}\,K^{-1}$) is not well known, because different climate models produce different values. In addition, part of the radiative forcing denoted ΔQ (W/m^2) is transported to the deeper ocean. The statement of heat balance of the lower atmosphere and upper ocean at the current time (now) is therefore that the heat flux into the deeper ocean is equal to the radiative forcing ΔF minus the radiative cooling,

$$\Delta Q_{now} = \Delta F_{now} - \lambda_{LW} \Delta T_{now}. \tag{3.1}$$

Consider a doubling ($2\times$) of the CO_2 concentration, leading to a radiative forcing of $\Delta F_{2\times}$. After the climate system has had time to warm and equilibrate to the new radiative forcing, the ocean absorbs no additional net heat (it is in a new, warmer, steady state and the net heat flux into the ocean must therefore vanish), and we have $0 = \Delta F_{2\times} - \lambda_{LW} \Delta T_{2\times}$. Our objective is to use observational constraints to calculate the anticipated final equilibrium warming for double CO_2, $\Delta T_{2\times}$. From the above, we have

$$\Delta T_{2\times} = \frac{\Delta F_{2\times}}{\lambda_{LW}}$$

as well as

$$\lambda_{LW} = \frac{\Delta F_{now} - \Delta Q_{now}}{\Delta T_{now}},$$

which may be combined to give an equilibrium warming of

$$\Delta T_{2\times} = \frac{\Delta F_{2\times} \Delta T_{\text{now}}}{\Delta F_{\text{now}} - \Delta Q_{\text{now}}}. \tag{3.2}$$

The present surface warming (ΔT_{now}) and current heat flux into the deep ocean (ΔQ_{now}) appearing on the RHS can be estimated from present-day observations, while the radiative forcings at present and for a doubling of CO_2 (ΔF, $\Delta F_{\times 2}$) can be estimated reliably from radiation models. Thus, the RHS is known, and this can be used to estimate the desired equilibrium climate sensitivity on the LHS without relying on complex and somewhat uncertain climate models (see this chapter's workshop). The result is, not surprisingly, in the same range predicted by these models.

3.1.2 Transient climate sensitivity

The deeper ocean absorbs some of the heat due to the excess radiative forcing, and its large heat capacity means that it warms up very slowly. Because the relatively cool deep ocean continuously mixes with the more rapidly warming surface ocean, it slows down the surface warming. This means that the currently observed surface warming, for example, reflects only part of the potential warming given the present greenhouse gas concentration; this section attempts to explain and quantify this effect.

Consider the response of the surface temperature (including the upper ocean) and the deeper ocean temperature to an abrupt doubling of the CO_2 concentration. The following equations describe the temperature evolution governed by the heat budget (per unit area) of the lower atmosphere and upper ocean, whose temperature perturbation due to the increase in greenhouse gases is denoted $\Delta T_{\text{surface}}$, and of the deeper ocean, whose temperature perturbation is denoted ΔT_{deep},

$$C_{\text{surface}} \frac{d\Delta T_{\text{surface}}}{dt} = \Delta F_{2\times} - \lambda_{LW} \Delta T_{\text{surface}} - \gamma \left(\Delta T_{\text{surface}} - \Delta T_{\text{deep}} \right)$$

$$C_{\text{deep}} \frac{d\Delta T_{\text{deep}}}{dt} = \gamma \left(\Delta T_{\text{surface}} - \Delta T_{\text{deep}} \right). \tag{3.3}$$

The first equation is the perturbation heat budget per unit area for the atmosphere and upper ocean due to the increase in greenhouse gases; the second, for the deeper ocean. $C_{surface}$ and C_{deep} are the heat capacities per unit area of the combined atmosphere and ocean surface and of the deep ocean, corresponding-ly. The left-hand side of both equations is therefore the rate of change of the heat content of the atmosphere / surface ocean and of the deep ocean, per unit area. The RHS represents heat fluxes responsible for this rate of change. The term $\gamma \left(\Delta T_{surface} - \Delta T_{deep} \right)$ represents the slow heat exchange between the deep ocean and the upper ocean, which was represented by ΔQ in equa-tion (3.1). The parameter γ is in units of heat capacity per unit area over a timescale, $(J\,K^{-1}\,m^{-2})\,s^{-1}$, where the (long) timescale is that of the slow heat exchange between the deep and surface ocean. Note that this term has opposite signs in the two budget equations, so that in a warming scenario it represents a source of energy for the deep ocean and a sink for the surface. The first equa-tion generalizes the equilibrium balance assumed previously in equation (3.1) with two differences: we now take into account the time rate of change of the surface temperature (on the LHS), and we consider the response to an *abrupt* CO_2 doubling, hence the use of $\Delta F_{2\times}$ in equation (3.3) rather than ΔF_{now}.

As for parameter values, for a CO_2 doubling, detailed calculations show that the radiative forcing is $\Delta F_{2\times} = 4$ W/m^2, and if the equilibrium climate sen-sitivity is assumed to be $3\,°C$, then $\lambda_{LW} = (4 \text{ W/m}^2)/(3\,°C)$. $C_{surface}$, which represents the combined heat capacity of the atmosphere and upper 50 m of the ocean, may be approximated as $C_{surface} \approx \rho_w c_p\, 50$ m; and the heat capacity of the deeper ocean is given by $C_{deep} = \rho_w c_p H$, where H is the ocean depth, $c_p = 4005\,J\,K^{-1}\,kg^{-1}$ is the specific heat capacity of seawater (heat required to raise the temperature of 1 kg of seawater by 1 K, in $J\,K^{-1}\,kg^{-1}$), and $\rho_w = 1024$ kg m^{-3} is the density of seawater. The parameter γ is set to one, and the second equation in (3.3) implies that the timescale over which the tempera-ture of the deep ocean is modified by the mixing term is γ / C_{deep}, or about 500 yr.

In a steady state, when the LHS of equation (3.3) vanishes, the second equa-tion gives $\Delta T_{surface} = \Delta T_{deep}$. This indicates that the deep ocean and surface ocean warmed to the same degree and are now in equilibrium. The first equa-tion then gives $\Delta T_{surface} = \Delta F_{2\times}/\lambda_{LW}$, as in section 3.1.1. Thus, the ocean depth and heat capacity do not enter the steady state solution and do not affect the equilibrium response to radiative forcing. However, in the short term, the ocean heat capacity can make a big difference, as demonstrated by Figure 3.3.

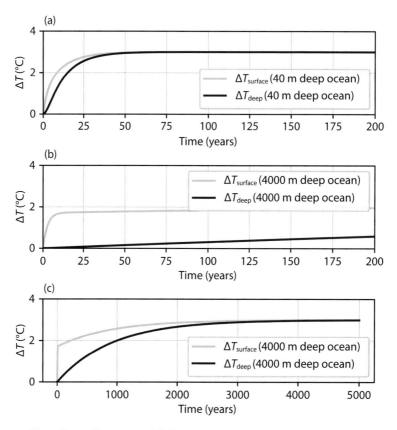

Figure 3.3: Transient climate sensitivity.
The temperature anomalies of the upper ocean and of the deep ocean, as a function of time in a scenario of instantaneous CO_2 doubling. (a) An artificial case assuming the subsurface ocean is only 40 m deep. (b) A more realistic scenario, assuming an ocean depth of 4000 m and showing only the first 200 yr of adjustment to an abrupt doubling of CO_2. (c) Same scenario as in (b), and showing the full period of adjustment.

The upper panel shows that in the case of a shallow ocean (where the deeper ocean is assumed to be only 40 m deep), both the ocean and atmosphere adjust to the new radiative forcing within about 50 yr. For the case in which the deep ocean depth is assumed to be more realistic (4000 m), the middle panel shows the initial adjustment, showing that the atmosphere adjusts quickly but only partially and that the deep ocean is very slow to respond. The lower panel shows the longer-term adjustment, demonstrating that the atmospheric adjustment is indeed delayed significantly and is only completed when the deep ocean has completed its warming after thousands of years. The warming prior to full equilibration, which takes into account the delayed response due to the heat capacity of the deep ocean, is referred to as the *transient* climate sensitivity.

One may conclude that while we have seen a warming of about a degree so far, this is a result of the ocean's heat capacity slowing the atmospheric warming rather than being the result of a low climate sensitivity to CO_2 increase.

3.2 POLAR AMPLIFICATION

The significantly stronger warming seen in the Arctic in both the warming that has occurred so far and in that anticipated in future projections (Figure 3.2) is referred to as *polar amplification* or *Arctic amplification*. This amplification means that the high latitudes are especially sensitive to global warming, with a range of potential implications from Greenland melting and sea level rise to Arctic wildlife to effects on remote mid-latitude weather and more. The polar/Arctic amplification is due to three main factors, which are reviewed in this section.

The albedo feedback

The melting of high-latitude snow and ice in a warmer climate, over both land and ocean, leads to a decrease in albedo, an increase in absorbed solar radiation, and therefore further warming. This is an important positive feedback leading to the warming of the high latitudes where snow and ice are currently widespread, and therefore to polar amplification. But the reduced high-latitude albedo is not the only responsible mechanism nor necessarily the strongest one leading to polar amplification.

The Planck feedback

Consider a simple equation balancing incoming radiation flux F with an outgoing black-body LW radiation, $F = \epsilon \sigma T^4$, where the temperature T is in Kelvin. If a radiative forcing ΔF is now added to the LHS, we expect the temperature to increase by ΔT, leading to a new balance,

$$F + \Delta F = \epsilon \sigma (T + \Delta T)^4. \tag{3.4}$$

A quick calculation using this equation shows, for example, that if $T = 240K$ ($-33\,°C$), an increase in radiative forcing of $\Delta F = 4\,W/m^2$ leads to about twice

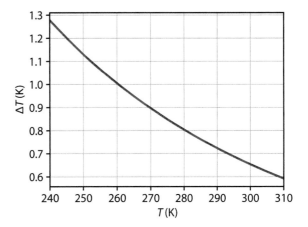

Figure 3.4: The Planck feedback.
The warming expected due to an increase in radiative forcing of 4 W/m² based on equation (3.5) with $\epsilon = 1$.

as large an increase ΔT than if $T = 300K$ ($27\,^\circ$C), letting the longwave emissivity be $\epsilon = 1$ for simplicity. Thus, the same radiative forcing leads to a larger temperature response if the base temperature is colder. This *Planck feedback* contributes to Arctic amplification. Equation (3.4) may be simplified and linearized using the fact that $(\Delta T)/T \ll 1$, so that higher powers of this ratio may be neglected. Expanding $(T + \Delta T)^4 = T^4(1 + (\Delta T)/T)^4$ and dropping all terms with powers of $(\Delta T)/T$ higher than one, we find $\Delta F \approx 4\epsilon\sigma T^3 \Delta T$, or

$$\Delta T \approx \Delta F/(4\epsilon\sigma T^3). \tag{3.5}$$

This expression makes it even clearer that the temperature response ΔT to radiative forcing ΔF is stronger at low temperatures T, as shown in Figure 3.4. The above discussion treated the radiative forcing ΔF as constant in latitude in order to isolate the effects of the Planck feedback on polar amplification, although in reality it may be latitude-dependent as well.

The tropical lapse rate feedback

The atmospheric lapse rate, dT/dz, where T is the atmospheric temperature and z the altitude in meters, represents the rate of cooling with altitude and varies between about -5 and -9.8 K/km. The lapse rate itself, as well as its change due to warming, depends on the surface temperature and on the atmospheric moisture, among other factors. This dependence leads to a different change to the lapse rate as a result of surface warming in the tropics versus high latitudes, a difference that contributes to polar amplification.

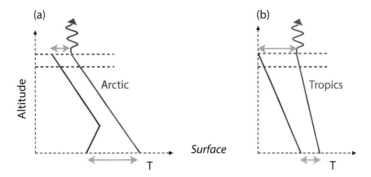

Figure 3.5: Schematics of the Arctic (a) and tropical (b) lapse rate feedbacks.
Solid blue (red) lines show the temperature profiles before (after) warming. Dashed blue (red) lines show the emission level before (after) the warming (see section 2.1.3). The green double arrows show the warmings at the surface and at the emission level.

Starting with the tropics, consider a parcel of moist air adiabatically rising (that is, the air parcel rises without exchanging mass or heat with its environment) with some specified surface temperature and moisture content. As the parcel rises, it encounters lower atmospheric pressure, expands and cools, and as a result experiences some condensation as well. Its temperature profile with height is known as the moist adiabatic profile, and its rate of cooling with height is the moist adiabatic lapse rate. A warming of the surface temperature (say, due to an increase in greenhouse gas concentration) changes the moist adiabatic temperature profile such that there is an even larger warming at high altitudes, as explained below. This reduction in the difference between the high altitude and surface temperatures implies that the tropical lapse rate becomes less negative with a warming of the surface temperature.

Given that the tropical atmospheric temperature as a function of altitude tends to follow the moist adiabatic profile, we expect that in a warmer climate, as the surface warms in the tropics, the higher troposphere there will warm even more. The stronger high-altitude warming leads to stronger radiative cooling from the upper troposphere to outer space via outgoing longwave radiation than would have been expected given the magnitude of the surface warming (see schematic Figure 3.5b). This implies that the change in tropical lapse rate is a negative feedback: the surface warms, and the increased cooling to outer space more than compensates for this warming, thus reducing the tropical surface warming.

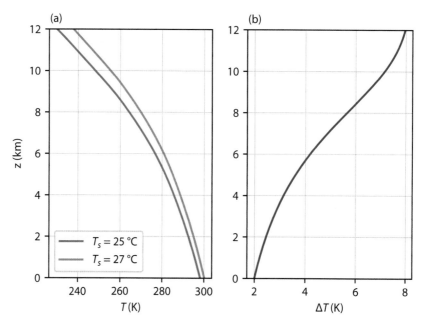

Figure 3.6: Tropical lapse rate feedback.
(a) Temperature profiles of two surface air parcels starting with a relative humidity of 100% and two different surface temperatures and rising adiabatically in the atmosphere. (b) The difference in temperature between the two profiles, showing the enhanced upper atmosphere warming of the parcel that starts with a slightly warmer surface temperature.

The weakening (in absolute value) of the tropical lapse rate with surface warming that is behind this negative feedback can be calculated using energy conservation for moist air parcels (further discussed in chapter 7) and can be qualitatively understood as follows. Consider two air parcels at the surface, one initially warmer than the other, and assume for simplicity that both are at saturation with respect to water vapor (having relative humidity of 100%). Due to the Clausius-Clapeyron relation (Box 2.1), the warmer parcel contains more moisture initially. As the parcel rises and encounters lower pressures, it expands and therefore cools adiabatically. The cooling leads to condensation and therefore to latent heat release. The parcel that started out warmer and moister will, as a result, experience more latent heat release and therefore more heating as it rises. This heating means a smaller rate of cooling as the parcel rises, and thus a less negative lapse rate. This implies that a warmer surface temperature leads to an even larger warming aloft. Figure 3.6 shows two such temperature profiles and their difference, calculated by considering the energy conservation of the rising moist air parcels, as explained using the concept of moist static energy and by solving the corresponding equations shown in section 7.2.

The high-latitude lapse rate response to warming

At high latitudes, the warming response to an increased greenhouse gas concentration tends to be surface-enhanced. As a result, a given surface warming is accompanied by a smaller warming in the upper troposphere. This leads to a smaller increase in outgoing LW radiative cooling to outer space than would have been expected from the surface warming, and this acts as a *positive* feedback on the surface warming. This enhanced surface warming and thus the positive Arctic lapse rate feedback result to a large degree from the fact that the winter Arctic surface in the present climate, especially over sea ice and snow, is very cold, and the near-surface winter-time atmospheric temperature *increases* with height. This is referred to as a surface *inversion* (blue line in Figure 3.5a). A surface–enhanced warming, as shown by the red line in Figure 3.5a, reduces the near-surface inversion strength (or eliminates the inversion), resulting in a more significant warming at the surface than at high altitudes.

The weakening of the inversion in a warmer climate can occur for several reasons. First, such an inversion typically occurs in the present climate over sea ice in the Arctic, so that a melting of the sea ice exposes the much warmer surface ocean and eliminates the inversion. A second possible mechanism in a warmer climate involves the response of low clouds to warming. As the ocean warms, so does the air over the ocean, which becomes moister as a result of the Clausius-Clapeyron relation. The moister air travels from the ocean over the cold high-latitude continents (e.g., Canada, northern United States) during winter-time and cools as a result. The moisture condenses and leads to the formation of clouds in the lower atmosphere. Because the solar radiation during winter is weak at high latitudes, the cloud albedo effect is minimal. At the same time, the low clouds absorb the LW radiation emitted by the surface and re-emit it downward and upward (section 2.1.2), thus leading to a strong cloud greenhouse effect. This leads to an enhanced warming of the lower atmosphere and thus to the weakening or elimination of the inversion and to the Arctic lapse rate feedback.

The combined response of the negative lapse rate feedback at low latitudes and positive lapse rate feedback at high latitudes leads to enhanced Arctic warming and reduced tropical warming, contributing to polar amplification.

3.3 "HIATUS" PERIODS

The globally averaged surface temperature record over the past century and a half shows a period of what seems to be temporary cooling (1940–1970 or so) and another period of seemingly paused warming (1998–2013). There is a significant body of research work looking at the mechanisms behind the temperature trends of these periods, especially the latter one. Yet it is useful to remember that natural climate variability leads to oscillations in the globally averaged temperature, and when these oscillations are superimposed on a warming trend, the two combine to give what seem to be "hiatus" periods. This is demonstrated in Figure 3.7, where the right panel shows the observed record for the surface temperature with the two "hiatus" periods marked by horizontal bars, and the left panel shows the globally averaged temperature in a control run (driven with a fixed preindustrial CO_2 concentration) of a climate model (green) and the same time series with an added linear trend (red). The control run record plus a linear trend produces what seem to be "hiatus" periods,

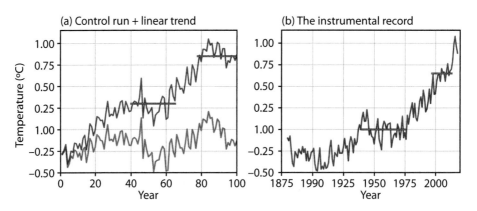

Figure 3.7: "Hiatus" periods.
(a) Globally averaged surface temperature anomaly from a control (CO_2 fixed at preindustrial concentration) run of a climate model (green), and the same temperature record with a linear trend added (red). (b) The instrumental record of globally averaged surface temperature anomaly. Both panels show "hiatus" periods marked by gray horizontal bars.

supporting the idea that such periods are to be expected in a global warming scenario. Furthermore, observations show there was no "hiatus" in the warming of the oceans over 1998–2013. Given that 90% of the excess heat trapped in the climate system due to the increase in greenhouse gas concentration ends up in the subsurface ocean, the "hiatus" in surface warming makes a little dent in the overall warming trend in the climate system.

3.4 STRATOSPHERIC COOLING

The distinguishing element of the stratosphere is its warming as a function of altitude, as opposed to the cooling as a function of altitude in the troposphere (Figure 3.8a). This is a result of absorption of SW radiation by the ozone layer in the stratosphere. Thus, unlike the troposphere, whose main source of radiative heating is LW radiation from the surface, the stratosphere is also heated by absorbed SW radiation. This leads to the very different stratospheric response to an increase in CO_2 concentration. Figures 3.8a,b show the zonally averaged temperature change (°C) in an RCP8.5 scenario during the 21st century as a function of height (a) and height and latitude (b), showing that while the troposphere warms, the stratosphere cools significantly.

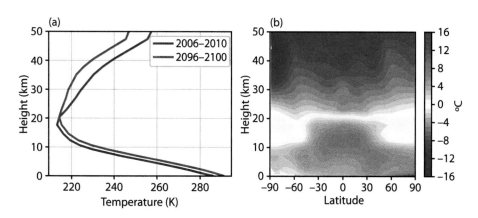

Figure 3.8: Stratospheric cooling.
(a) Mid-latitude (30°N–50°N) zonally averaged temperature profiles for an RCP8.5 projection at the beginning and end of the 21st century. (b) The zonally averaged atmospheric temperature response during the 21st century to the RCP8.5 scenario, showing a tropospheric warming and a stratospheric cooling.

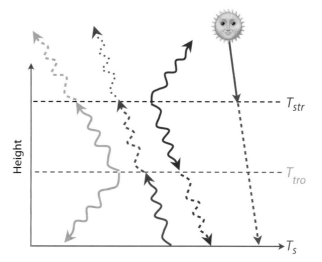

Figure 3.9: Understanding strato-spheric cooling.
An energy balance model including a stratospheric layer, used to explain the stratospheric cooling in a global warming scenario.

To understand the response of the stratosphere, we add a third layer to our two-level energy balance model from section 2.1.2, such that T_s is the surface temperature, T_{tro} the tropospheric temperature, and T_{str} the newly added strato-spheric temperature, as shown in the schematic Figure 3.9.

We assume that a fraction β_{str} of the SW solar radiation is absorbed by the stratosphere. The energy balances of the surface, troposphere, and stratosphere can then be written with the energy sources on the LHS and sinks on the RHS as

$$(1 - \beta_{str}) \frac{1}{4} S_0 + \epsilon_{tro} \sigma T_{tro}^4 + (1 - \epsilon_{tro}) \epsilon_{str} \sigma T_{str}^4 = \sigma T_s^4$$

$$\epsilon_{tro} \sigma T_s^4 + \epsilon_{tro} \epsilon_{str} \sigma T_{str}^4 = 2 \epsilon_{tro} \sigma T_{tro}^4$$

$$\beta_{str} \frac{1}{4} S_0 + \epsilon_{str} (1 - \epsilon_{tro}) \sigma T_s^4 + \epsilon_{str} \epsilon_{tro} \sigma T_{tro}^4 = 2 \epsilon_{str} \sigma T_{str}^4. \qquad (3.6)$$

In these equations, each term represents a heat flux in W/m^2, and all temper-atures are in Kelvin. The terms involving $\frac{1}{4} S_0$ represent incoming SW solar radiation, as explained in section 2.1.1; ϵ_{tro} is the tropospheric emissivity and ϵ_{str} the stratospheric. The different terms may be interpreted using the schematic arrows in Figure 3.9. For example, the solid red arrow shows the LW radiation from the surface at a rate of σT_s^4. When it encounters absorption in the tropo-sphere, the red arrow changes from solid to dashed, and a heat flux $\epsilon_{tro} \sigma T_s^4$ is deposited there. Then, the remainder surface emission flux, $(1 - \epsilon_{tro}) \sigma T_s^4$, con-tinues upward, as represented by the dashed red arrow. The line changes again in the stratosphere from dashed to dotted, corresponding to the deposition of a heat flux $\epsilon_{str} (1 - \epsilon_{tro}) \sigma T_s^4$ there as well, with the remainder continuing to outer space (dotted red arrow). These equations can be written as an easily solved linear set of equations for the temperatures raised to the fourth power,

$$
\begin{pmatrix}
\sigma & -\epsilon_{tro}\sigma & -(1 - \epsilon_{tro})\epsilon_{str}\sigma \\
-\epsilon_{tro}\sigma & 2\epsilon_{tro}\sigma & -\epsilon_{tro}\epsilon_{str}\sigma \\
-(1 - \epsilon_{tro})\epsilon_{str}\sigma & -\epsilon_{str}\epsilon_{tro}\sigma & 2\epsilon_{str}\sigma
\end{pmatrix}
\begin{pmatrix}
T_s^4 \\
T_{tro}^4 \\
T_{str}^4
\end{pmatrix}
$$

$$
=
\begin{pmatrix}
(1 - \beta_{str}) \frac{1}{4} S_0 \\
0 \\
\beta_{str} \frac{1}{4} S_0
\end{pmatrix}.
$$

As a crude qualitative numerical example, consider $\beta_{str} = 0.05$ (the stratosphere, in the range of 1–100 hPa, absorbs about 5–15 W/m² depending on latitude and season), $\epsilon_{tro} = 0.55$, $\epsilon_{str} = 0.1$, and, for an increase in CO_2, let $\epsilon_{tro} = 0.6$, $\epsilon_{str} = 0.15$. The solution before the increase in CO_2 is $T_s = 302$ K, $T_{tro} = 257.4$ K, and $T_{str} = 260.1$ K. The solution with the emissivity values representing the increased CO_2 concentration is $T_s = 306$ K, $T_{tro} = 261.8$ K, and $T_{str} = 254$ K, showing the warming of the surface and the troposphere and the cooling of the stratosphere, as expected.

To understand the reason for this cooling, consider first the two-level energy balance model of equation (2.3) in section 2.1.2. As CO_2 concentration increases, the tropospheric emissivity (denoted ϵ there) increases, increasing both the source and sink terms in the tropospheric heat budget (ϵ drops out of that equation, appearing on both sides). For the surface equation, the increased atmospheric emissivity increases the heat source of downward LW and thus leads to surface warming. The atmosphere then responds to the increase in emission from the warmer surface and warms up as well (while this is a helpful idealization in the current context, remember the refinements discussed in section 2.1.3 involving the lapse rate and emission height). Adding a stratospheric layer as the third equation in (3.6) does not qualitatively change this response of the troposphere and surface (first two equations). However, for the stratosphere, because of the presence of the SW absorption heating term $(\beta_{str} \frac{1}{4} S_0)$, the stratospheric emissivity ϵ_{str} does not drop out of the third equation in (3.6). The LW cooling on the RHS is balanced by both LW absorption from the lower atmosphere and surface and by SW absorption. When the stratospheric emissivity increases, say by P%, due to the increase in greenhouse gas concentration, the cooling term on the RHS increases by this percentage. However, in the source terms on the LHS, only the two LW radiative heating terms increase by P%, while the SW heating does not increase with the increased stratospheric emissivity. As a result, the sum of the heating terms on the LHS increases by a smaller fraction than the radiative cooling term $2\epsilon_{str}\sigma\, T_{str}^4$ on the RHS, leading to a net cooling of the stratosphere with increasing CO_2.

3.4.1 Detection and attribution

A successful *detection* of a signal due to anthropogenic climate change (say, in global mean surface temperature, stratospheric temperature, or Arctic sea ice area) requires showing that the observed change is inconsistent with internal/natural variability. An *attribution* of observed events (say, of the strong Arctic sea ice melting in 2012 or the Siberian heat wave of 2020 or the California forest fires of 2020) to climate change due to greenhouse gas increase specifically requires further showing that the observed change is *only* consistent with the suggested forcing scenario (e.g., natural variability plus increase in greenhouse forcing) and is *not consistent* with alternative, plausible explanations of recent climate change due to either anthropogenic or natural sources (e.g., a change in ozone concentration, aerosols, or land use or a response to volcanoes).

A stratospheric cooling was predicted as early as 1967 to be an expected consequence of CO_2 increase, and combined with a warming of the troposphere, it is a clear signature of CO_2-induced climate change. This is therefore an important example of the attribution of observed atmospheric temperature changes to the increase in greenhouse gas rather than to other anthropogenic effects (e.g., ozone depletion) or natural causes (e.g., volcanoes).

3.5 WORKSHOP

A Jupyter notebook with the workshop and corresponding data file are available; see https://press.princeton.edu/global-warming-science.

1. **Characterize the warming, historical and future projections:**

   (a) Hockey-stick curve: Plot the global-mean, annual-mean temperature anomaly as a function of time, for observations, continued by the RCP8.5 and RCP2.6 scenarios.

   (b) Spatial distribution of warming: Contour the RCP8.5 2100 temperature minus the preindustrial temperature at 1850 as a function of latitude and longitude.

   (c) Polar amplification: Plot the zonally averaged warming versus latitude for the RCP8.5 scenario.

2. **"Hiatus" periods:**

   (a) Plot the given global-mean, annual-mean climate model control-run temperature variability simulated at preindustrial CO_2, with and without an added linear trend of $1\,°C$/century. Observe the "hiatus" periods formed.

   (b) Plot the observed global temperature anomaly for 1970–1998, fit a linear trend, and extend the plot of the linear fit through the last year of data. Then plot the data after 1998 to see whether they preferentially fall below the projected trend line. Comment on whether the data during the supposed post-1998 "hiatus" period support the idea that global warming ceased in 1998.

3. **Equilibrium climate sensitivity:** Assume the current radiative forcing anomaly due to anthropogenic emissions is $2.3\,W/m^2$ and the current global-mean surface warming is $1\,°C$. Assume also that 50% of the radiative forcing heat flux currently goes into the deep ocean. Estimate the equilibrium warming at a double-preindustrial CO_2 concentration (2×280 ppm), assuming it leads to a radiative forcing of $4\,W/m^2$.

4. **Transient climate sensitivity and the role of the ocean:** Solve equations (3.3) and plot ΔT_{deep} and $\Delta T_{\text{surface}}$ for $t = 0, \ldots, 5000$ yr, for $H = 40$ m and for $H = 4000$ m. Explain the role of the ocean's heat capacity in the response to global warming. Show directly from the equations that the steady state solution is consistent with the plotted results.

5. **Polar amplification:**

(a) Planck feedback: (i) Consider $F = \epsilon \sigma T^4$, letting $\epsilon = 0.6$, calculate the warming ΔT due to an increase in radiative forcing of $\Delta F = 4$ W/m^2, if $T = -10\,°C$, and then if $T = +30\,°C$. (*Hint:* It is easier to calculate ΔF from T and ΔT for different values of ΔT.) (ii) Calculate dF/dT for these two temperatures.

(b) *Optional extra credit:* Tropical lapse rate feedback: Calculate and plot the temperature profiles of saturated air parcels being raised in the tropical atmosphere, with initial surface temperatures of $25\,°C$ and $27\,°C$, and the difference between the two profiles (see section 7.2 for help). Discuss your results.

(c) *Optional extra credit:* Lapse rate feedback: Assume the radiative forcing due to greenhouse gases is $\Delta F = 8.5$ W/m^2 and that the radiation balance at the emission level is given by the linearized expression $\Delta F = \lambda_{LW} \Delta T_e$, where ΔT_e is the warming at the emission level. Further assume that the lapse rate feedbacks are such that the tropical (Arctic) warming at the emission height is a factor X ($1/X$) times that at the surface. What should X and λ_{LW} be to explain a surface Arctic amplification of 10 K, such that the averaged surface warming in the two areas is 6 K?

6. **Stratospheric cooling:**

(a) Contour the zonally averaged temperature change over the 21st century in the RCP8.5 scenario as a function of latitude and height in the troposphere and stratosphere; plot the profiles of the temperature averaged zonally and between latitudes of 30°N and 50°N before and after a projected RCP8.5 warming.

(b) Calculate the expected tropospheric warming and stratospheric cooling as the tropospheric and stratospheric emissivities change from $\epsilon_{tro} = 0.55$ and $\epsilon_{str} = 0.1$ to $\epsilon_{tro} = 0.62$ and $\epsilon_{str} = 0.17$, for an increase in CO_2, following section 3.4.

(c) Calculate each term in the stratospheric energy balance before and after the CO_2 increase (that is, before and after the increase in the emissivities in the third equation in 3.6). Show that if the stratospheric temperature does not change in response to the emissivity change, the increase in the RHS term representing outgoing LW from the stratosphere (both up and down) due to the increase in stratospheric emissivity is larger than the increase in the incoming radiation on the LHS. Deduce that the stratosphere must cool to maintain its heat balance.

7. **Guiding questions to be addressed in your report:**

(a) Describe the extent of surface warming so far, and its distribution as a function of latitude, for land versus ocean and for winter versus summer.

(b) Explain the latitudinal dependence; what are the implications for high-latitude countries? For Greenland melting and thus for other regions?

(c) Explain how future warming is evaluated from present-day data in addition to the use of climate models.

(d) If a doubling of CO_2 from 280 ppm to 560 leads to a, say, 3 K warming, what warming do you expect at an equivalent CO_2 mixing ratio of 450 ppm? Explain why the warming by 2020 was only about 1.1 K.

(e) Should we be concerned about global warming given that it is said to have stopped recently for 15 yr? Will it stop again?

SEA LEVEL

Xiaoting Yang and E.T.

Key concepts

- The historical record and future projections
  - Exiting from the Little Ice Age versus anthropogenic global warming
  - Decadal variability
  - Global versus regional
  - Future projections
- Global mean sea level change
  - Thermal expansion
  - Glacier and ice sheet mass balance
  - Land water storage

- Regional sea level change
  - Wind stress
  - Atmospheric sea level pressure
  - Ocean circulation
  - Land erosion
  - Gravitational effects

Global mean sea level (GMSL) has increased by about 35 cm over the past 150 yr (Figure 4.1). While some of this increase is likely related to the natural warming due to the exit from the Little Ice Age (extending roughly from 1300 to about 1850), the signature of anthropogenic climate change on sea level is clearly seen. Recent events, from the flooding of Manhattan during Hurricane Sandy in 2012 to the flooding of Venice, Italy, in 2019 that involved a 1.8 m local sea level rise, brought sea level to the forefront of the news. We will see that

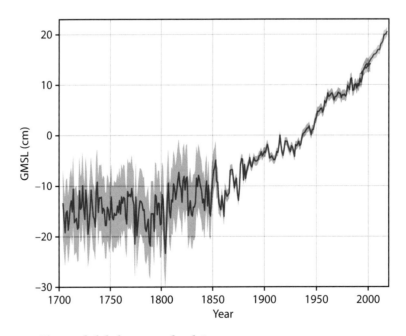

Figure 4.1: Observed global mean sea level rise.
Observed global mean sea level anomaly (defined as the deviation from a specified reference value) since 1700 (blue curve) and the estimated uncertainty (light-blue shading). The recent satellite record and its uncertainty range are similarly shown in red and orange.

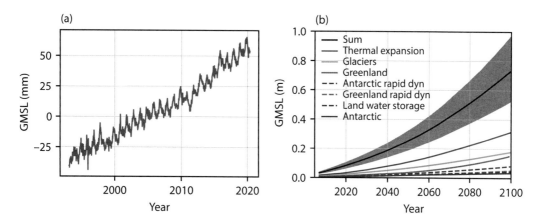

Figure 4.2: Observed and projected global mean sea level change.
(a) Satellite record since 1992. (b) Future projections under RCP8.5, including the contribution of different processes, are discussed in this chapter. The black curve and gray shading show the projected total sea level rise and its uncertainty.

both natural processes and anthropogenic climate change can contribute to such events. Additionally, evidence from past warm climates suggests the potential for a significant sea level rise in response to fairly modest greenhouse warming (Box 4.1). Sea level changes involve a wide range of regional and global processes, and we attempt here to understand and quantify the main ones.

Sea level rises and falls on many timescales. For example, GMSL rose 130 m since the last glacial maximum 21 kyr ago, due to the melting of ice sheets that covered the high-latitude continents during the last ice age (Box 10.1). More recently, global mean sea level has been rising at a rate of 3.5 mm/yr and seems to have accelerated in recent decades (Figures 4.1, 4.2a). Future projections suggest that GMSL will continue rising, by up to 1 m by the end of the century under the Representative Concentration Pathway 8.5 scenario (Figure 4.2b).

Starting with some definitions, the *global mean sea level* is a measure of the total water volume in the oceans. *Relative sea level (RSL)* is the height of the sea surface above the solid ocean bottom at a particular location. *Steric* sea level reflects the variation of the ocean volume due to ocean water density changes alone, induced by warming, for example.

In the following sections we first examine GMSL changes due to processes such as glacier and ice sheet melting, acceleration of ice flow from Greenland and Antarctica, expansion of warming ocean water, and more (section 4.1). Local

RSL due to wind or ocean current changes, local land uplift or sinking of the coast, coastal land erosion, and the gravitational effect of melting in Greenland and Antarctica will be examined in section 4.2.

4.1 GLOBAL MEAN SEA LEVEL CHANGES

Processes that affect GMSL include, although are not limited to, (1) thermal expansion, (2) glacier and ice sheet mass balance, and (3) land water storage.

4.1.1 Thermal expansion

As discussed in section 3.1, about 90% of the excess radiative forcing due to the increase in greenhouse gas concentration has so far been stored in the warming oceans. This causes seawater to expand, contributing to GMSL rise. Fresh water freezes at $0\,°C$, and its density as a function of temperature peaks at $4\,°C$. However, for salty ocean water, the freezing temperature is $-1.8\,°C$, and its density decreases with temperature above this freezing point. The typical vertical ocean temperature profile is characterized by a few different regimes (Box 6.1). The upper 25–50 m or so is typically well-mixed by wind and waves and is termed the *mixed layer*, where temperature is nearly uniform as a function of depth. Below the mixed layer, the temperature decreases within the *main thermocline* down to about 1 km depth, below which the temperature is low and nearly uniform. When the atmosphere warms, the mixed layer quickly warms within months, the above-thermocline ocean warms up over decades, while the deep water may take hundreds to thousands of years to warm.

Based on the observed warming of the oceans for the period 1971–2010 and their deduced thermal expansion, the rate of GMSL rise due to the expansion of ocean water from the surface to 700 m depth is estimated at about 0.6 mm/yr, and the rate becomes 0.8 mm/yr if the deeper ocean is included. By the end of the century, warming according to the RCP8.5 scenario will lead to a sea level rise due to thermal expansion of over 30 cm, a significant fraction of the anticipated rise (Figure 4.2b).

A back-of-the-envelope calculation

Consider a warming of the upper ocean and the associated sea level rise due to the thermal expansion of seawater. Initially, d_0 is the depth of the water column experiencing warming, its temperature is denoted T_0, and its initial density is $\rho(T_0)$. After experiencing a warming ΔT, the water has a temperature $T = T_0 + \Delta T$, an expanded depth $d = d_o + \Delta d$, and a reduced density $\rho(T_0 + \Delta T)$. Given that the mass of ocean water is not changed by the warming, we can write an equation stating that the mass (per unit area) before warming is equal to that after the warming,

$$m = d \times \rho(T_0 + \Delta T) = d_0 \times \rho(T_0),$$

which implies that $d = d_o \rho(T_0)/\rho(T_0 + \Delta T)$, or

$$\Delta d = d - d_0 = d_0 \left(\frac{\rho(T_0)}{\rho(T_0 + \Delta T)} - 1 \right)$$

$$= d_0 \frac{\rho(T_0) - \rho(T_0 + \Delta T)}{\rho(T_0 + \Delta T)}.$$

To a good approximation, $\rho(T_0 + \Delta T)$ in the denominator may be replaced by a constant reference (averaged) density ρ_0, giving

$$\Delta d = d_0 \frac{\Delta \rho}{\rho(T_0 + \Delta T)} \approx d_0 \frac{\Delta \rho}{\rho_0}, \tag{4.1}$$

so that sea level rise is proportional to d_0, the depth of penetration of the warming.

As an example, consider a uniform warming of $\Delta T = 3\,^\circ\mathrm{C}$ corresponding to a mid-range IPCC surface warming due to CO_2 doubling. Assume this warming penetrated to a depth of 500 m, and let the temperature before the warming be $T_0 = 16\,^\circ\mathrm{C}$. The ocean water density depends on temperature, as mentioned above, but also on the salt concentration (salinity, measured in parts per thousand, ppt, roughly representing kg salt per metric ton of seawater). Let the density before warming be $\rho(16\,^\circ\mathrm{C}, 35\,\mathrm{ppt}) = 1025.75\,\mathrm{kg/m^3}$ and the density after warming be $\rho(19\,^\circ\mathrm{C}, 35\,\mathrm{ppt}) = 1025.022\,\mathrm{kg/m^3}$. Note that these density values justify replacing $\rho(T_0 + \Delta T)$ with $\rho_0 = 1025\,\mathrm{kg/m^3}$ in the denominator of equation (4.1) with a negligible effect on the accuracy of the calculated sea level rise. Using the above expressions, we find a sea level rise of $\Delta d = 35$ cm, similar to the IPCC estimate of sea level rise due to thermal expansion in the RCP8.5 scenario (red line in Figure 4.2b).

Calculating GMSL change from ocean warming

More generally, we need to take into account that ocean warming varies in both depth and the horizontal direction and that the expansion of seawater depends on its temperature. Consider the *equation of state* relating ocean water density to temperature. A good first approximation to this equation of state is that the density at a constant salinity S_0 changes linearly with temperature according to

$$\rho(T, S_0) = \rho_0(1 - \alpha_T(T - T_0)), \qquad \alpha_T = -\frac{1}{\rho_0}\frac{\partial \rho}{\partial T}\bigg|_{T=T_0},$$

where T_0 is the temperature prior to the warming, T the temperature after the warming, and α_T the *thermal expansion coefficient,* itself a function of temperature and evaluated here at T_0. This is a linearization of the equation of state using Taylor expansion. To calculate the change to GMSL, let $\Delta T(x, y, z) = T - T_0$ be the warming as a function of horizontal location and depth, divide the ocean into layers of infinitesimal thickness dz, and apply the above mass balance to each vertical layer in the ocean with dz replacing d_0 in equation (4.1). Integrating over all layers from the ocean bottom at $z = -H$ to the surface at $z = 0$ and over all horizontal locations (x, y), we find that the increase in ocean volume is

$$\Delta V = \int dx \int dy \int_{-H}^{0} \frac{1}{\rho_0} \Delta \rho(x, y, z)dz.$$

The change to GMSL is this change in volume divided by the ocean area A. Using the fact that $\Delta \rho = \rho(T) - \rho(T + \Delta T) \approx \rho_0 \alpha_T \Delta T(x, y, z)$, we find the change in GMSL to be

$$\Delta \text{GMSL} = \frac{1}{A}\int dx \int dy \int_{-H}^{0} \frac{1}{\rho_0} \Delta \rho(x, y, z)dz$$

$$= \frac{1}{A}\int dx \int dy \int_{-H}^{0} \alpha_T \Delta T(x, y, z)\, dz.$$

The dependence of the expansion coefficient on temperature is significant. For example, assuming a salinity of 35 ppt, $\alpha_T(2\,^\circ\text{C}) = 0.78 \cdot 10^{-4}\,^\circ\text{C}^{-1}$, while $\alpha_T(10\,^\circ\text{C}) = 1.6 \cdot 10^{-4}\,^\circ\text{C}^{-1}$, reflecting the sensitivity of the density to the temperature: $\rho(2\,^\circ\text{C}) - \rho(3\,^\circ\text{C}) = 0.09\,\text{kg/m}^3$, while $\rho(10\,^\circ\text{C}) - \rho(11\,^\circ\text{C}) =$

0.18 kg/m^3. As a result, warming the already warm low-latitude surface ocean leads to a larger contribution to sea level rise than warming the colder high-latitude or deep waters.

Climate Background Box 4.1
Past warm climates

An important perspective on possible future climate response to warming, including of sea level, can be obtained by examining past warm climates. Panel a in the figure below shows a proxy of the deep ocean temperature over the past 60 Myr, showing the gradual cooling since the very warm *hot-house* climate of the Eocene (56–33.9 Myr). Past ocean temperatures may be deduced from extracted deep sea sediment cores that contain the shells of plankton deposited at the ocean bottom for millions of years. The isotopic composition of these shells (e.g., $\delta^{18}O$, which is a function of the ratio between the concentrations of oxygen isotopes ^{16}O and ^{18}O) depends on the temperature in which the plankton grew, thus recording past temperatures. The vertical axis in panel a is a function of the oxygen isotopic composition of plankton shells from deep ocean sediments, converted to a Celsius-like scale for convenience. There is fossil evidence of frost-intolerant species of plants and animals existing during the Eocene in areas such as present-day northern North America, where temperatures currently often reach $-40\,^{\circ}C$ in winter. This suggests that surface winter temperatures at that time did not drop below freezing, even in high-latitude continental areas away from the moderating effects of the ocean. Similarly, deep ocean temperatures may have been as warm as $12\,^{\circ}C$, as opposed to around $0\,^{\circ}C$ at present, indicating that the high-latitude areas where deep ocean water sinks during winter to fill the abyssal ocean (Box 6.1) were also warm. During the Eocene, there was no significant ice over Antarctica nor Greenland, so that sea levels were more than 70 m higher than at present. Note the large scatter of dots in the past 2–3 Myr, an expression of the large variability due to the glacial cycles (Box 10.1).

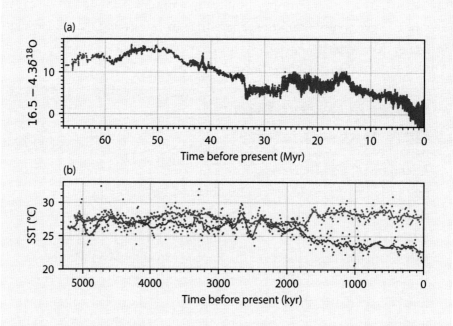

(a)

16.5 − 4.3δ¹⁸O

Time before present (Myr)

(b)

SST (°C)

Time before present (kyr)

The last time atmospheric CO_2 concentration is believed to have been around 400 ppm, similar to that at present, was the Pliocene period, 5.3–2.6 Myr, which is therefore considered the nearest past analog of a future global warming. Global mean surface temperature was some 2–3 °C warmer than at present, sea level has been estimated to have been up to 20 m above present levels, the Arctic was possibly 8 °C or more warmer, and the equatorial Pacific may have been in a permanent El Niño state, as shown in panel b. This panel shows reconstructed surface ocean temperature time series from the east and west Pacific, showing them to be separated by some 4 °C all the way to about 2 Myr ago, and to be much closer prior to that. During present-day El Niño events (Box 8.1), as the east Pacific warms, the east and west Pacific have a similar temperature, as seen in the Pliocene record during 2–5 Myr, hence the interpretation of a permanent El Niño. Additionally, during the Pliocene some mid-latitude ocean upwelling sites (such as off the coast of California) were some 10 °C warmer than today, suggesting that the upwelling was disrupted. These upwelling sites are currently among the richest fisheries in the world ocean, because the upwelling brings nutrients to the surface, which supports the growth of plankton and attracts fish. This Pliocene climate may represent an equilibrium response to present-day CO_2 levels, toward which our climate might be progressing assuming current CO_2 levels are sustained for a long time, as discussed in section 3.1. These past warm climates are still not well understood and not well simulated by state-of-the-art climate models, indicating the need to improve our understanding and modeling of warm climate dynamics.

4.1.2 Ice sheets and mountain glaciers

Greenland and Antarctica

Ice sheets gain mass via snow accumulation and lose it via *ablation*, which is the combination of all processes that lead to ice loss. This includes the net surface mass balance (snow accumulation minus surface ice melting/sublimation), ice flow toward the ocean in ice streams, iceberg calving into the ocean, and more. For the period of 1993 to 2010, the contribution to global mean sea level rise from Greenland was 0.4 mm/yr, and the contribution from Antarctica was 0.27 mm/yr, although it is currently difficult to estimate how much of this signal is due to natural variability and how much is due to anthropogenic climate change (as a reminder, GMSL is currently increasing by about 3.5 mm/yr). Note that the RCP8.5 projections in Figure 4.2b include a contribution due to a possible acceleration of ice flow from Greenland and Antarctica. This contribution is referred to as *rapid dynamics*, and is distinct from changes to surface accumulation or melting. A warmer climate is expected to lead to more snow accumulation over Antarctica (reducing the rate of sea level rise) due to the increase in atmospheric water vapor with warming (Clausius-Clapeyron, Box 2.1). At the same time, (uncertain) future projections anticipate increased ice flow, which is projected to dominate the expected increased snow accumulation by the end of the century. The expected total contribution of Antarctica to GMSL is therefore positive, although less than the expected contribution of rapid dynamics alone.

For much more on the contribution of Greenland and Antarctica, and on the processes and uncertainty involved, see chapter 10.

Mountain glaciers

Mountain glaciers have been retreating for about 150 yr, contributing to GMSL rise (for more on mountain glaciers, see chapter 11). Some of the melting occurred before any significant anthropogenic warming and likely represents delayed climate response to exiting from the Little Ice Age. There has been an acceleration in mountain glacier melting in recent decades reflecting the effect of anthropogenic warming, and the US National Snow and Ice Data Center has reported, for example, that the mass balance of 41 monitored glaciers has become more negative from 1980 to 2012. The contribution to GMSL from glaciers excluding Greenland and Antarctica was 0.76 mm/yr during 1993 to 2010.

4.1.3 Land water storage

Fresh water accumulation in groundwater, lakes, and reservoirs makes a negative contribution to GMSL. Similarly, the pumping and use of groundwater removes land-trapped water, some of which then evaporates and rains over the oceans and contributes to GMSL rise. The increase in groundwater withdrawal is slowly over-weighting the negative contribution made by water impoundment by the constructions of dams, for example. For the period of 1993 to 2010, the contribution to GMSL from reductions in land water storage was 0.38 mm/yr.

4.1.4 Detection of anthropogenic climate change in GMSL

Figure 4.1 leaves little doubt that there is a strong trend over the past century that is very distinct from the natural variability prior to 1850 or so. Thus, one might conclude that the signs of climate change are clearly detected in the GMSL record. While we will later examine a few quantitative statistical approaches to this detection issue (e.g., sections 12.3 in the context of droughts and 9.2 in the

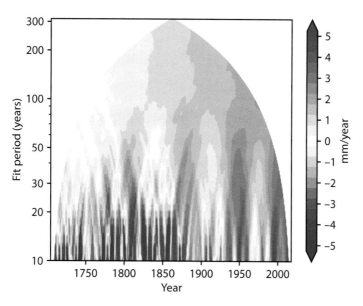

Figure 4.3: Detecting climate change in sea level rise.
Shown are the rates of sea level rise in mm/year as a function of year (horizontal axis) and averaging (fit) period (vertical).

context of sea ice), here we instead use a visualization approach, as formal statistics do not seem necessary in this case. Figure 4.3 shows the sea level trend as a function of year and averaging period. To obtain this figure, we consider an averaging period, say 30 yr. We then fit a linear trend at any point in the record to the segment from the 30 yr period centered at that point and record the slope of the fitted line. Proceeding to do so for averaging periods from 10 to 300 yr, we obtain the shown contour plot showing the rate of sea level change (line slope) as a function of year and fit period.

Consider first the bottom of the plot, for averaging periods of 10 to 15 yr or so. At the beginning of the record, the trends can be either positive or negative, alternating rapidly and representing natural variability and perhaps also noise due to larger errors in the less well observed earlier record. But later, especially after 1990, the trends are uniformly positive, where the trend clearly masks any variability signal. At averaging periods longer than about 30 yr, the right side of the plot is uniformly red, again showing that rise in GMSL dominates the variability.

4.2 REGIONAL SEA LEVEL CHANGES

Regional sea level changes can be much faster and larger than GMSL changes, and the associated local sea level drop or rise can be very dramatic. The main factors that influence regional patterns of relative sea level change include atmosphere-ocean interaction (wind stress, atmospheric pressure loading, and ocean current changes), spatial differences in the rate of ocean warming, land changes due to coastal erosion, isostatic rebound of a continent in response to the melting of past land ice sheets, and the gravitational effect due to current melting of ice sheets.

4.2.1 Atmosphere-ocean interaction

Atmospheric pressure loading

Regional sea level adjusts to atmospheric pressure changes. Atmospheric sea level pressure (SLP) exerts a vertical force on the ocean surface; as a result, a high SLP leads to a low sea surface height (SSH), and a low SLP to a high SSH. For example, a low sea level pressure in the center of a hurricane—say, 50 hPa below normal SLP—can lead to a sea level rise of about 50 cm (roughly 1 cm per 1 hPa). This effect is distinct from the wind-driven storm surge caused by storms, as discussed below. In a warmer climate, the large-scale sea level atmospheric pressure is expected to change due to changes in atmospheric circulation, increase in the subtropics and mid-latitudes, and decrease at higher latitudes, especially over the Arctic. In the RCP8.5 scenario this leads to a fairly modest, although consistent among different climate models, sea level rise of about 2.5 cm in the Arctic and in the Southern Ocean near Antarctica and a smaller sea level decrease spread over lower latitudes. Smaller-scale changes in SLP involved in local weather conditions can be significantly larger.

To calculate the sea level response to SLP changes, consider an initial SLP that does not vary horizontally, consistent with a flat SSH and a motionless ocean (Figure 4.4a). A positive SLP perturbation Δp, applied only regionally (Figure 4.4b), pushes water away, causing a sea level drop under the higher pressure and sea level rise elsewhere. Horizontal pressure differences in the ocean lead to water movement, and for the new ocean state to be at equilibrium, the

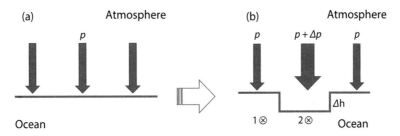

Figure 4.4: Atmospheric pressure loading.
How a change to sea level atmosphere pressure (SLP) from spatially uniform on the left to spatially variable on the right affects regional sea level.

water pressure within the ocean should remain horizontally uniform. The pressure below the ocean surface is determined by both the sea level atmospheric pressure and the sea surface height, which reflects the weight of ocean water above a given point. For the ocean to be at rest, the changes to ocean pressure due to the sea level difference should balance the SLP perturbation. The pressure difference between two horizontal locations within the ocean (e.g., those marked $1\otimes$ and $2\otimes$ in Figure 4.4b) due to an SSH difference Δh between these locations is the weight of water (per unit area) due to this difference, equal to $\rho g \Delta h$, where ρ is the ocean water density and g is gravity. The condition that the total subsurface pressure within the ocean under the SLP perturbation is equal to the total subsurface ocean pressure away from there is therefore $\Delta p + \rho g \Delta h = 0$, allowing us to calculate the SSH response Δh. Writing $\Delta h = -\Delta p/(\rho g)$, with $\rho \approx 1025 \, \text{kg m}^{-3}$ and $g = 9.8 \, \text{m}^2\text{s}^{-2}$, we find that an SLP signal of 50 hPa (5000 Pa) leads to a sea level change of 0.5 m, as mentioned above. This simple calculation does not take into account the Coriolis force due to the Earth rotation, discussed shortly.

Wind stress

Wind changes due to weather or climate variability, or due to climate change, can lead to a significant regional sea level response. A near-surface wind of velocity U (m/s), typically measured at 10 m above the ocean surface, results in a horizontal force per unit area (*stress*, in units of N/m^2) on the ocean surface. Consider an ocean channel, crudely representing the Adriatic Sea, for example, of length L_1 and width L_2. Figure 4.5 shows the wind stress τ pushing

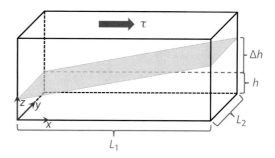

Figure 4.5: Wind-driven sea level change.
Wind stress drives a sea surface height slope and affects sea level height.

the water in a channel to the right, creating a sea level height difference Δh across the length of the channel. The higher sea level on the right implies that the ocean pressure is higher there as well, exerting a push on the water toward the left. When the two forces due to the wind and pressure exactly balance, the water level is in a new equilibrium, allowing us to calculate the equilibrium SSH response to the wind.

The wind force over the ocean is given by the *bulk formula* $\tau = \rho_{air} C_D |U|U$, where the force is quadratic in the wind speed, and the use of absolute value of the wind speed times the wind speed itself guarantees that the sign (direction) of the force is the same as that of the wind. Here, surface air density is $\rho_{air} = 1.2$ kg/m^3, and C_D is a nondimensional *drag coefficient*, with a typical value of 0.0013. The total wind force over the channel surface is $F_{wind} = \tau L_1 L_2$. The pressure at a depth z in the ocean may be calculated as the weight per unit area of water over that level, $p(z) = \rho g(h - z)$, where h is the local sea surface height, and z is the vertical coordinate, increasing upward and set to zero at the bottom, so that $h - z$ is the height of the water column above a depth z. Assuming that the pressure at a given depth is the result of the water weight above that point is referred to as the *hydrostatic balance*. Integrating over the cross-section area perpendicular to the wind direction, in the (y, z) plane, the horizontal pressure force on this section is $L_2 \int_0^h \rho g(h - z)dz = \frac{1}{2}L_2\rho gh^2$. The net horizontal pressure force on the channel in Figure 4.5 is the pressure force from the left (where the ocean depth is h) minus that from the right (where it is $h + \Delta h$), or

$$F_{\text{pressure}} = \left[\frac{1}{2}\rho gh^2 - \frac{1}{2}\rho g\,(h + \Delta h)^2\right]L_2 \approx -\rho gh\Delta hL_2. \qquad (4.2)$$

In the last equation we have neglected terms proportional to the square of the height difference, $(\Delta h)^2$, because they are smaller. Using the horizontal force balance $F_{\text{wind}} + F_{\text{pressure}} = 0$, we have $\tau L_1 L_2 = \rho g h \Delta h L_2$, so that the sea level change across the channel is estimated as

$$\Delta h = \tau L_1 / (g \rho \bar{h}),$$

where $\bar{h}$ is the averaged ocean depth. Thus, a large SSH response can result from a small mean depth $\bar{h}$, a large wind stress, or a long distance L_1 over which the wind blows. This expression shows that a persistent wind due to a weather event over the Adriatic Sea could lead to a large sea level rise, as observed in Venice, Italy, in 2019 and in 1966. It is very difficult to attribute such individual strong wind events to ACC; what can typically be done is to attempt to estimate to what degree the probability of a given wind event was enhanced due to ACC, as we demonstrate below in other contexts. Therefore, in general, the detection of and attribution of regional sea level events is more challenging than for the GMSL record (discussed in section 4.1.4).

Storm and tidal surges

A strong storm, such as a tropical cyclone (chapter 8), can affect regional sea level via both the pressure loading and wind stress effects discussed above, as well as via the forcing of large-amplitude ocean waves. When combined with local high tides that happen to occur at the same time, these can lead to a very dramatic local sea level change, coastal flooding, and significant damage. Local infrastructure meant to address storm and tidal surges can be overwhelmed when these effects are combined with longer-term changes due to GMSL rise, longer-term regional sea level rise associated with climate change, and changing storm strength due to climate change.

Ocean currents and the Coriolis force

On a large enough scale of tens of kilometers and above, the Coriolis force enters the ocean dynamics by which pressure loading and wind stress affect sea level response. The Coriolis force acts on every object moving relative to the rotating Earth, to the right of the motion in the Northern Hemisphere and to the left in the Southern Hemisphere. Due to the Coriolis force, ocean currents such as the Gulf Stream are accompanied by an SSH difference across the current (there is about a meter SSH difference between the two sides of the Gulf Stream!). A

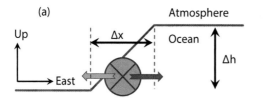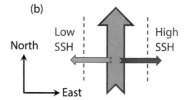

Figure 4.6: The Coriolis force and sea level.
Sea surface height variations across the northward flowing Gulf Stream, showing the pressure force from high to low sea level (green arrows) and Coriolis force to the right of the current (red arrows). (a) A vertical east-west section (current is denoted by the blue symbol ⊗ of an arrow pointing into the page). (b) A horizontal schematic with the Gulf Stream shown by the blue arrow.

---

change to ocean currents will change the height difference across the current and can therefore also lead to a regional SSH change. This may be quantified by analyzing the force balance dominating ocean currents.

The dominant force balance for practically all major ocean currents (and large-scale atmospheric winds and weather systems) is between the Coriolis force and the pressure force. These forces generally balance each other, such that the net force is zero; as a result the current is at a steady state and does not accelerate. Consider a schematic east-west section across the northward flowing Gulf Stream (drawn as flowing into the page and denoted by the blue ⊗ symbol in the vertical section of Figure 4.6a and by the blue arrow in the plane view of Figure 4.6b). The SSH to the east of the Gulf Stream is about 1 m higher than that west of the stream. This leads to a westward pressure force from high sea level to low sea level (green arrow). This force is balanced by the Coriolis force: The Gulf Stream flows northward, and in the Northern Hemisphere this leads to a Coriolis force to the right of the flow—that is, eastward (red arrow)—that exactly balances the pressure force.

To quantify this discussion, consider an ocean current flowing northward (in the y direction) associated with an east-west sea level difference Δh (in the x direction). The horizontal pressure force due to a height difference Δh across the current, per unit distance along the current, is derived above (eqn 4.2) as

$$F_{\text{pressure}} \approx -\rho g h \Delta h. \tag{4.3}$$

The acceleration tendency experienced due to the Coriolis force by an ocean current moving northward at a velocity v (m/s) is given by fv, where $f = 2\Omega \sin \theta$ is the Coriolis parameter (units of s^{-1}), the rotation rate of the Earth is given by $\Omega = 2\pi/\text{day} = 0.0000727$ (s^{-1}), and θ is the latitude. The Coriolis force is this acceleration times the mass of the relevant ocean section (per unit distance along the current), $\Delta x h \rho$, where Δx is the width of the current and h the ocean depth. That the Coriolis force and the pressure force sum to zero in an equilibrium implies that $0 = (\Delta x h \rho)fv - \rho g h \Delta h$, or $-fv = -g\Delta h/\Delta x$. Writing $\Delta h/\Delta x = \partial h/\partial x$, we find that the force balance of a northward flowing ocean current such as the Gulf Stream is

$$-fv = -g\frac{\partial h}{\partial x}.$$

The left-hand side represents the eastward acceleration driven by the Coriolis force on the northward flowing Gulf Stream, while the right-hand side represents the westward acceleration due to the east-west SSH gradient leading to a pressure force across the stream.

A similar balance between the pressure and Coriolis forces holds for other ocean currents and is referred to as *geostrophy*. If a current velocity changes, so would the corresponding sea surface height gradient and as a result also the local sea level near the current, possibly affecting sea level at nearby coastal locations. Ocean currents may respond to climate change, for example, due to a wind change driven by an atmospheric response to a warmer ocean temperature. Substituting order-of-magnitude values for the Gulf Stream, of a velocity $v = 1$ m/s, width $\Delta x = 100$ km, Coriolis parameter $f = 2\Omega \sin(30°) = 7.27 \times 10^{-5}$ s^{-1}, and $g = 9.8$ m/s^2, we find the height difference across the Gulf Stream to be about $\Delta h = fv\Delta x/g = 74$ cm. If the current weakens by 10% due to surface wind changes associated with global warming, this will reduce the height difference across the current by a corresponding fraction, possibly leading to a sea level rise along the coast of North America.

For the sake of completeness, we note that if a current is flowing in the east-west x direction at a velocity u, its geostrophic balance can be similarly derived to be $fu = -g\partial h/\partial y$. These expressions for geostrophy do not take into account the important effect of change in ocean water density with depth, which is beyond our scope here. For the sake of simplicity, our discussion of the effects of wind stress on sea level height did not take into account the Coriolis force, although in reality the Coriolis force does play a role in how sea surface height responds to wind changes. Finally, we note again that geostrophy is also the force balance of major wind patterns in the atmosphere, from weather systems to atmospheric jet streams.

4.2.2 Land changes

Coastal erosion and land subsidence

Coastal erosion is the loss or displacement of the upper part of the land surface adjacent to the coastline due to the action of waves, currents, tides, wind-driven surges, waterborne ice, precipitation runoff, and other impacts of storms. This may lead to the retreat of the coastline landward and to an increase in relative sea level over the eroded coast. The connection of specific erosion events to global warming, due for example to an expected strengthening of tropical storms (chapter 8), may be indirect and therefore uncertain. Dramatic recent land erosion events occurred in Kutubdia, an offshore island in Bangladesh in 2012, due to tidal waves, high winds, and strong rainfall, in New Jersey and Long Island in response to Hurricane Sandy, and elsewhere. In addition, land subsidence (lowering) in or adjacent to an ocean-covered area similarly leads to the increase in relative sea level rise. Subsidence can be caused by the over-pumping of groundwater, by the compaction of the ground due to centuries of building, and by long-term changes due to plate tectonics. Venice, Italy, is estimated to have subsided by about 120 mm in the 20th century, contributing about half of its observed sea level rise. Of course, natural processes such as sediment transport by rivers or by waves and tides can also lead to the *buildup* of coastal areas and to the decline of relative sea level. Such natural processes may be disrupted by coastal building and by the elimination of natural vegetation, adding to the effects of climate change.

Post-glacial rebound and isostatic adjustment

The lingering rising of the solid Earth after the melting of a land ice sheet or glacier can last thousands of years; therefore, the influence of the melting of the large ice sheets of the last glacial maximum (21 kyr; see Box 10.1) still affects sea level changes observed today. When a large ice mass forms on a continent, its weight presses the continental surface down, leading to a subsurface flow within the upper mantle away from the glacier until a new equilibrium is reached. Given that ice density ($900 \ kg/m^3$) is about a third of that of the solid Earth, a third of the height of the ice is submerged below the Earth surface at equilibrium (Archimedes' law). The response of the continent to the presence or melting of ice sheets is referred to as isostatic adjustment. When the ice melts away, the continent rebounds, and there is a return subsurface flow within the upper mantle. As a result, the continent rebounds near the location of the melting ice sheet and sinks away from there. This leads to a downward trend in the relative sea level close to the location of the ice sheet and to an opposite trend away from it. These sinking land trends can lead to a *regional* sea level increase of about a millimeter per year along parts of the Atlantic coast of the United States, and need to be taken into account when interpreting local sea level records for the signal of global warming. This effect is distinct from the more straightforward effect on the *global* sea level of the melting of the ice sheet and the flow of its melt water to the ocean.

4.2.3 Gravitational effects: sea level fingerprints of melting

A sea level gravitational fingerprint of melting is a spatial pattern of the response of sea level to ice sheet melting due to gravitational effects. Consider a melting of the Greenland ice sheet. Because of the addition of water mass to the ocean, GMSL will rise, *but the rise is not uniform* as a function of horizontal location. Currently, the gravitational attraction by the mass of ice over Greenland means that sea level is increasing toward Greenland. The reduction of ice mass due to melting results in less gravitational attraction of ocean water toward Greenland, which would tend to *reduce* sea level near Greenland. Since sea level near Greenland is reduced, sea level elsewhere is expected to rise more than the GMSL rise expected due to the melting. This gravitational effect leading to local reduction of sea level in response to Greenland melting extends some 2000 km away from Greenland and therefore influences the sea level rise near parts of Northern Europe. Figure 4.7 shows the fingerprints of melting in Greenland and west Antarctica.

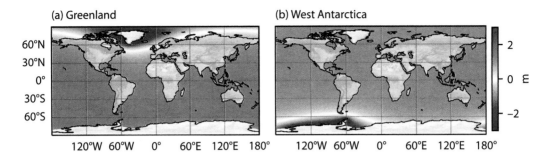

Figure 4.7: Gravitational effects.
Sea level fingerprint of melting in (a) Greenland, (b) West Antarctica, showing distribution of sea level rise in meters, where the GMSL rise is 1 m in both cases.

In parallel, the reduction of ice weight will make the underlying continent adjust upward elastically over a short period and further adjust slowly due to isostatic adjustment over a longer timescale (section 4.2.2). The loss of ocean water near Greenland due to the weakening of the gravitational force will make the sea floor adjust upward as well. On the other hand, the addition of water away from Greenland will push the sea floor there downward. The resulting spatial pattern of sea level change is the fingerprint of Greenland melting and is different from the fingerprint of Antarctic melting.

Due to these processes, a rate of Greenland melting that corresponds to a 1 mm/yr rise in GMSL is expected to lead to sea level drop within 2000 km of Greenland and at the shore of Greenland itself can lead to a *drop* of up to 10 mm/yr, 10 times larger than the global rate of opposite sign!

4.3 WORKSHOP

A Jupyter notebook with the workshop and corresponding data file are available; see https://press.princeton.edu/global-warming-science.

1. **Characterizing sea level rise:**

   (a) GMSL: (i) Plot a time series of the observed globally averaged sea level anomaly since 1850, followed by the RCP8.5 projection to year 2100, and a quadratic polynomial fit to the observed time series. (ii) Discuss how and why the *rate* of the sea level rise in the two periods changes over this period. (iii) Based on your quadratic fit, and assuming the fit and thus the sea level rise *acceleration* remain the same, what would sea level be at 2100? Is the assumption of a constant acceleration justified?

   (b) Spatial structure of sea level rise: Contour the estimated spatial structure of sea level rise across the observed period (1850–2005) and for the projected RCP8.5 period (2006–2100). Explain your results; in particular discuss the signal around the Southern Ocean.

2. **Temperature, density and sea level rise:**

   (a) Plot ocean water density as a function of temperature for $T = -2, \ldots, 30\,°C$. Plot the expansion coefficient α_T ($°C^{-1}$) for the same temperature range. Discuss the implications for sea level rise of the dependence of the expansion coefficient on temperature.

   (b) Assuming warming is exponential in depth, $\Delta T(z) = 4e^{z/500}$ with z zero at the ocean surface and increasing upward, calculate the expected sea level rise. Assume the equation of state is $\rho = \rho_0 [1 - \alpha_T (T - T_0)]$, in which $\rho_0 = 1025$ kg/m^3, $\alpha_T = 1.668 \times 10^{-4}\,°C^{-1}$, and $T_0 = 10\,°C$. What is the expected sea level rise if the exponential decay scale is 1000 m instead of 500 m? Explain.

   (c) ***Optional extra credit:*** Given a certain amount of heat (in J) added to the climate system, would it cause more sea level rise if it melts land ice or leads to expansion of seawater? Your answer should take into account the initial temperatures of the ice and of the ocean. Contour the ratio of the two possible contributions to sea level rise as a function of these two temperatures.

3. **Warming-driven sea level change patterns:** Given the estimated three-dimensional temperature at 1850 and the CMIP5 RCP8.5 projection at 2100, calculate and plot the local contribution to sea level rise as a function of longitude and latitude, and calculate the expected GMSL rise. Assume that salinity is a constant at 35 ppt, considering only the effects of warming.

4. **Wind forcing, sea level pressure, and the Coriolis force:**

   (a) Consider a storm in the Adriatic Sea north of Monte Gargano, Italy, with wind blowing along the longer dimension of the sea toward Venice at 30 m/s. Calculate the sea level rise at the far north end of the sea.

   (b) Calculate the sea level change expected in the center of a category 4 hurricane whose center pressure is 920 to 944 hPa.

   (c) Calculate the change in sea level difference across the Gulf Stream at a latitude 30°N in response to a weakening of the current by 10%. Assume the original flow was 1 m/s and that the width of the Gulf Stream there is 50 km.

5. **Sea-level gravitational fingerprint:** Consider the sea level fingerprints in Figure 4.7. Explain these sea level change patterns.

6. *Optional extra credit:* **Decadal variability versus climate change.** Reproduce Figure 4.3 using the GMSL record since year 1700, but fitting a quadratic polynomial and contouring the acceleration rather than slope of the fitted line. *Discuss:* What are the implications of your analysis to the interpretation of the sea level acceleration during the most recent 20 yr?

7. **Guiding questions to be addressed in your report:**

   (a) How and why is sea level changing along the Atlantic coast of the United States?

   (b) How is past GMSL rise calculated, and what are the uncertainties (touch on spatial patterns and observational coverage of ocean warming now versus two hundred years ago, wind changes, isostatic response to glacial melting, coastal erosion, gravitational fingerprint effects, and so on).

   (c) What is the projected GMSL rise by year 2100 based on RCP8.5? List the main processes responsible for the expected rise, quantify the contribution of each process, and describe the sources of uncertainty in each.

(d) Discuss the possible effect of anthropogenic global warming on storm surges and GMSL rise, and their interaction in specific regional sea level rise events.

(e) Discuss the expected effect of the projected GMSL rise by 2100 on New York City and Bangladesh. What are your recommendations in each case?

OCEAN ACIDIFICATION

Key concepts

- The ocean carbonate system, alkalinity, total CO_2, pH
- The effect of increasing atmospheric CO_2 on ocean acidity and on calcium carbonate dissolution
- Response to warming
- Long-term decline of anthropogenic CO_2

About a quarter of the CO_2 emitted by human activity has been absorbed by the world oceans, and another quarter by soils and the terrestrial biosphere, with the rest accumulating in the atmosphere. While this ocean absorption delays atmospheric warming, it leads to an acidification of the world oceans, referred to as "the other CO_2 problem." That the already observed non-negligible acidification of the world ocean will be followed by a significant

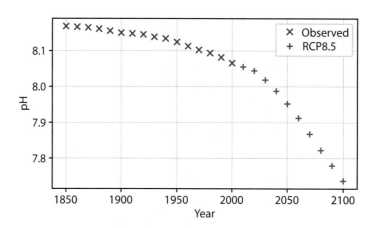

Figure 5.1: Ocean pH changes.
Observed globally averaged surface ocean pH from 1850 to 2000 (blue), and the projected pH to the end of the 21st century based on the RCP8.5 scenario (red).

further acidification is a robust prediction for a higher CO_2 world. The consequences for ocean biology are more difficult to predict, but it is difficult to imagine how the anticipated significant changes to ocean acidity would not lead to a disruption to ocean ecosystems. The acidity of the ocean is measured in terms of its pH, which is defined as the negative of the $\log_{10}$ of the *molar concentration* (moles per liter) of H^+ ions in seawater. Figure 5.1 shows the observed surface ocean pH since 1850, as well as that projected under the RCP8.5 scenario. The observed reduction in pH is from 8.16 to 8.06, which implies an increase of $100 \times (10^{-8.06} - 10^{-8.16})/10^{-8.16} = 25\%$ in the concentration of H^+ ions.

When CO_2 dissolves in seawater, it reacts with water to form carbonic acid, H_2CO_3, making the water more acidic (lower pH). This, as we will see, increases the concentration of bicarbonate (HCO_3^-) ions and decreases that of carbonate (CO_3^{2-}) ions. Multiple marine organisms deposit calcium carbonate ($CaCO_3$) structures as one of two minerals (crystal forms) known as aragonite and calcite, with the first being more soluble. This deposition becomes more difficult as the concentration of the carbonate ion decreases with decreasing pH. Among the marine organisms affected are single-celled phytoplankton such as coccolithophores and the common single-celled foraminifera, as well as pteropods, tiny (< 1 cm long) marine snails that are at the base of the oceanic food chain and whose shell is expected to begin to dissolve at acidity levels anticipated within this century, according to some high-emission scenarios. Furthermore, the increased acidification expected by year 2100 is suggested to lead to reduced olfactory functioning of fish and possibly disrupt their ability to detect prey and to avoid predators. Similarly, some studies suggest an effect of acidification on fish retinal function and vision.

The solubility of calcium carbonate depends on pH but also on temperature and pressure. Due to these dependencies, the deep ocean is *undersaturated* to the forms of calcium carbonate deposited by marine organisms, which means that it is already sufficiently acidic to dissolve these calcium carbonate structures. The upper ocean is *supersaturated* with respect to these forms; they therefore do not dissolve there. The transition zone between the dissolving deep ocean and the upper ocean has risen by 50–200 m in the past 150 yr due to the increase in atmospheric CO_2 concentration. Because cold water tends to be more undersaturated, the effects of ocean acidification are expected to be felt first in the colder higher latitudes and in the deep ocean.

As the ocean warms, its solubility to CO_2 decreases, reducing its ability to absorb anthropogenic CO_2 emissions and thus acting as a positive feedback on anthropogenic warming. Understanding and predicting this effect is important to our ability to predict warming later this century. Assuming anthropogenic emissions cease at some point, the emitted CO_2 will eventually be eliminated and the natural state recovered, although this can be expected to occur on a timescale of thousands to tens of thousands of years.

In the following, we attempt to understand calcium carbonate ($CaCO_3$) dissolution (section 5.1), describe the ocean carbonate system that controls this dissolution (section 5.2.1), understand this system by simplifying it to a set of reactions that can be solved explicitly (section 5.2.2), and then use this understanding to explore the response of pH and calcium carbonate dissolution to increasing atmospheric CO_2 concentrations (section 5.3.1). The mechanisms and consequences of decreasing CO_2 solubility in a warming ocean are discussed in section 5.3.2, and the restoring of the natural level of CO_2 over tens of thousands of years after emissions ceased is discussed in section 5.3.3.

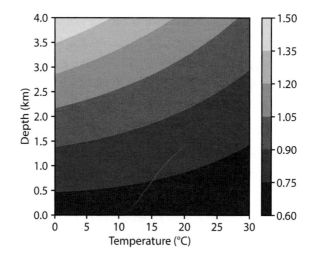

Figure 5.2: Solubility constant.
The calcium carbonate (specifically, aragonite) solubility product constant, K_{sp} (in $(mol/l)^2 \times 10^6$) as a function of temperature and depth (reflecting ocean pressure).

5.1 CALCIUM CARBONATE (CACO$_3$) DISSOLUTION

Many marine organisms produce calcium carbonate shells or skeletons in the form of calcite or aragonite and may be affected by a lower ocean pH. The equilibrium reaction between solid calcium carbonate and the ions it is composed of is written as

$$CaCO_3 \rightleftharpoons Ca^{2+} + CO_3^{2-},$$

and the equilibrium concentrations of the ions are related via the solubility product,

$$K_{sp} = \left[Ca^{2+}\right]\left[CO_3^{2-}\right], \tag{5.1}$$

where the brackets denote the molar concentration of a given ion. The solubility product constant itself is the equilibrium constant for a solid substance dissolving in an aqueous solution and represents the concentration levels at which a solute dissolves. The more soluble a substance is, the higher the K_{sp} value is. This constant for calcium carbonate depends fairly strongly on temperature and pressure (depth in the ocean), as shown in Figure 5.2. Given the solubility product, the potential for dissolving $CaCO_3$ is measured by the *saturation state* omega (Ω) defined as

$$\Omega = \frac{\left[Ca^{2+}\right]\left[CO_3^{2-}\right]}{K_{sp}}. \qquad (5.2)$$

When $\Omega < 1$, $CaCO_3$ tends to dissolve. As ocean pH decreases due to the increase in atmospheric CO_2, the carbonate ion concentration, $\left[CO_3^{2-}\right]$, decreases as well, making Ω smaller and leading toward more dissolution of calcium carbonate. In the following, we attempt to understand the link between atmospheric CO_2, pH, and CO_3^{2-} by formulating and solving the equations for the ocean carbonate system.

Climate Background Box 5.1
The carbon cycle

Carbon is exchanged between different reservoirs in the climate system, including in the deep ocean (which contains about 38,000 gigatons carbon, GtC), surface ocean (1000 GtC), atmosphere (750 GtC), biosphere (600 GtC), and soils (1600 GtC). The fluxes between these components involve a fast exchange of large fluxes between the land biosphere/soils and the atmosphere, due to photosynthesis and respiration, of about 120 GtC/yr (2 in the accompanying figure) and between the ocean biology and the atmosphere (90 GtC/yr (3)). In addition, the source of carbon in the climate system, and of atmospheric CO_2 in particular, is a slow release by volcanic activity, at a rate of 0.07 GtC/yr (1). There is also a slow sink due to the dissolution of CO_2 in rain (4), an interaction of the resulting solution with rocks and the formation of calcium carbonate that is dissolved in rivers (5), flows to the ocean, and is deposited in sediments there (6). On geological timescales, the ocean floor with the calcium carbonate sediments subducts under continental margins due to plate tectonics, and the carbon is recycled into the CO_2 emitted by volcanoes (7). Human emissions are currently about 10 GtC/yr, far outweighing the natural source due to volcanoes. With available fossil fuel reservoirs estimated at 5000 GtC, there is a potential for a significantly larger future disruption of the carbon cycle.

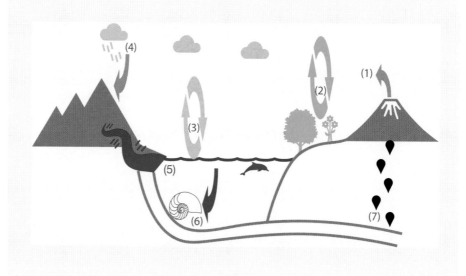

5.2 THE CARBONATE SYSTEM

Our objective is to find the relation between atmospheric CO_2, dissolved CO_2, CO_3^{2-}, ocean pH, and other related variables. We start by introducing the relevant chemical reactions that are part of the ocean carbonate system. Carbon dioxide is soluble in water, and its dissolution occurs in two steps. First, Henry's law states that atmospheric $CO_2(g)$ concentration is in equilibrium with dissolved $CO_2(aq)$,

$$CO_2(g) \rightleftharpoons CO_2(aq),$$

and then its reaction with water is given by

$$CO_2(aq) + H_2O \rightleftharpoons H_2CO_3 \text{ (carbonic acid)}.$$

Because it is difficult to distinguish between $CO_2(aq)$ and H_2CO_3, they are treated together as a single variable defined as

$$H_2CO_3^*(\equiv CO_2^*) \equiv CO_2(aq) + H_2CO_3. \tag{5.3}$$

In terms of this variable, Henry's law is

$$CO_2(g) \rightleftharpoons H_2CO_3^*. \tag{5.4}$$

Now, carbonic acid is a weak diprotic acid (diprotic acids are able to release two protons),

$$H_2CO_3^* \rightleftharpoons H^+ + HCO_3^- \tag{5.5}$$

$$HCO_3^- \rightleftharpoons H^+ + CO_3^{2-}. \tag{5.6}$$

Finally, water dissociation is given by

$$H_2O \rightleftharpoons H^+ + OH^-. \tag{5.7}$$

In order to solve for the six unknown concentrations of $CO_2(g)$, $H_2CO_3^*$, HCO_3^-, CO_3^{2-}, OH^-, H^+, we so far have only four reactions that will later be written as explicit equations: (5.4), (5.5), (5.6), and (5.7). We therefore need to specify two more constraints. One is mass conservation for the total number of carbon atoms, expressed via a quantity known as total dissolved inorganic carbon (DIC), also referred to as total CO_2 and denoted ΣCO_2 or C_T, which is conserved in the above reactions. The other constraint is the conservation of electric charge, which again must be satisfied by the above reactions and is expressed via a parameter called *alkalinity* (*Alk*). C_T and *Alk* are both measurable quantities for which one can write conservation equations affected by various sources and sinks, as well as by the movement and mixing of water masses. Once DIC and alkalinity are specified at a given location in the ocean, the carbonate system is completely determined (that is, there are the same number of unknowns and equations) and the concentrations of the different ions can be calculated. Consider these two constraints in some detail now.

Total CO_2

The number of moles carbon atoms per liter (MC/l) is given by the sum of the different species of the carbonate system,

$$DIC \equiv \Sigma CO_2 \equiv C_T = \left[H_2CO_3^*\right] + \left[CO_3^{2-}\right] + \left[HCO_3^-\right]$$

$$\approx \left[CO_3^{2-}\right] + \left[HCO_3^-\right],$$

where the partition in the ocean of the species appearing on the RHS of the first line is 1%, 10%, 90%, correspondingly (this partition is a strong function of pH), hence the approximation in the second line.

Alkalinity

The concept of alkalinity arises when one considers the charge balance of seawater. The net charge needs to be zero, which, when we take into account the major ions in the ocean, means that the number of positive charges minus the number of negative charges should vanish,

$$0 = \left([H^+] + [Na^+] + [K^+] + 2[Mg^{2+}] + 2[Ca^{2+}] \right)$$
$$- \left([HCO_3^-] + 2[CO_3^{2-}] + [OH^-] + [Cl^-] \right.$$
$$\left. + 2[SO_4^{2-}] + [NO_3^-] + [HBO_3^-] \right). \tag{5.8}$$

Now, we are interested in how some of these ion concentrations change with pH, CO_2, and other factors. For this purpose, it is useful to differentiate between strong bases and acids whose concentration does not change with pH and weaker ones that do change. For example, when NaCl dissolves in seawater, it separates completely into Na^+ and Cl^- regardless of the pH. However, in the dissociation of the weak acid HCO_3^-, $HCO_3^- \rightleftharpoons H^+ + CO_3^{2-}$, the concentrations of the ions on the RHS and LHS of this equilibrium vary with pH. Alkalinity is a measure of the charge balance due to these weak acids and bases. It is defined as the sum of negative ions that belong to weak acids or bases that change their dissociation with the ocean pH, minus the sum of positive ions that originate from such weak acids/bases. Separating the charge balance (eqn 5.8) into the parts due to the weak acids and bases (first line) and strong ones (second and third lines), we have

$$0 = \left([H^+] - [OH^-] - [HCO_3^-] - 2[CO_3^{2-}] - [HBO_3^-] \right)$$
$$+ \left([Na^+] + [K^+] + 2[Mg^{2+}] + 2[Ca^{2+}] - [Cl^-] \right.$$
$$\left. - 2[SO_4^{2-}] - [NO_3^-] \right).$$

Defining the alkalinity to be the negative of the first line, we have

$$Alk \equiv \left[HCO_3^-\right] + 2\left[CO_3^{2-}\right] + \left[HBO_3^-\right] - \left[H^+\right]$$

$$= \left[Na^+\right] + \left[K^+\right] + 2\left[Mg^{2+}\right] + 2\left[Ca^{2+}\right] - \left[Cl^-\right]$$

$$- 2\left[SO_4^{2-}\right] - \left[NO_3^-\right]. \tag{5.9}$$

Alkalinity is measured in units of *Equiv/l*, which is the sum of ion concentrations (e.g., in micromoles per liter), each multiplied by their charge. We will see below that the concentrations of H^+ and OH^- are very small, and we approximate the alkalinity by what is known as the *carbonate alkalinity*, as defined in the first line below, and then further approximate it as shown in the second line,

$$Alk_C \equiv \left[HCO_3^-\right] + 2\left[CO_3^{2-}\right] + \left[OH^-\right] - \left[H^+\right]$$

$$\approx \left[HCO_3^-\right] + 2\left[CO_3^{2-}\right]. \tag{5.10}$$

5.2.1 Carbonate system equations

To perform calculations, we write the above carbonate system reactions as the following set of equations using the equilibrium constants. The six unknowns are

$$\left[CO_2(g)\right], \left[H_2CO_3^*\right], \left[OH^-\right], \left[H^+\right], \left[HCO_3^-\right], \left[CO_3^{2-}\right], \tag{5.11}$$

where we remember that

$$H_2CO_3^*(\equiv CO_2^*) \equiv CO_2(aq) + H_2CO_3, \tag{5.12}$$

and the first four equations are

$$K_H \equiv K_0(T, S, p) = \frac{\left[H_2CO_3^*\right]}{\left[CO_2(g)\right]} \tag{5.13}$$

$$K_1(T, S, p) = \frac{\left[H^+\right]\left[HCO_3^-\right]}{\left[H_2CO_3^*\right]} \tag{5.14}$$

$$K_2(T, S, p) = \frac{[H^+][CO_3^{2-}]}{[HCO_3^-]} \tag{5.15}$$

$$K_w(T, S, p) = [H^+][OH^-]. \tag{5.16}$$

For typical values of these reaction coefficients on the LHS, which depend on temperature T, salinity S, and pressure p, see section 5.2.2. To close the system, we need two more equations, the definitions of carbonate alkalinity and total CO_2,

$$Alk_C = [HCO_3^-] + 2[CO_3^{2-}] + [OH^-] - [H^+] \tag{5.17}$$

$$C_T = [HCO_3^-] + [CO_3^{2-}] + [H_2CO_3^*]. \tag{5.18}$$

Given the values of Alk_C, C_T, K_H, K_1, K_2, and K_w, we can use the four carbonate system equations (5.13, 5.14, 5.15, 5.16) and the definitions of alkalinity and total CO_2 (5.17, 5.18) to solve for the six unknowns (5.11). This is a nonlinear system of equations, and it may be solved numerically. Note that instead of specifying Alk_C and C_T, we could have specified any two of the carbonate system variables, such as $CO_2(g)$ and Alk_C or C_T and pH. Once two are specified, the rest are calculated using the above equations.

The solution of the carbonate system as a function of pH for a constant alkalinity is shown in Figure 5.3. This figure was obtained by varying the DIC (C_T), specifying a fixed alkalinity, and solving for all variables for each value of C_T. The calculated pH is then used as the x-axis coordinate, as is customary for such plots. As a reminder, the current average surface ocean pH is about 8.1. Note that the carbonate ion (orange line), whose concentration controls the dissolution of calcium carbonate via the saturation state (eqn 5.2), decreases significantly for lower pH values, while the concentration of the bicarbonate ion (blue) increases. This solution provides the information regarding the response of the carbonate system as a function of pH required for us to calculate the ocean response to increased CO_2. The solution used to plot Figure 5.3 also includes the value of the atmospheric $CO_2(g)$ concentration that would have been in equilibrium with a volume of water with each particular value of the prescribed DIC. We use this to plot pH and the carbonate ion concentration as a function of $CO_2(g)$ in Figure 5.4, showing a significant reduction of the

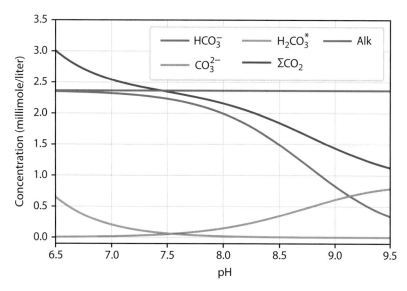

Figure 5.3: The solution of the carbonate system, showing the concentration of carbonate species as a function of pH for a fixed alkalinity.

carbonate ion concentration as the atmospheric $CO_2(g)$ increases. An even better understanding of the system can be developed by considering an approximate set of carbonate system equations that can be solved directly, as we do in the next section.

5.2.2 Approximate solution of the carbonate system

Consider an approximate solution to the carbonate system that allows us to better understand the response of the system to increased atmospheric CO_2 in a future global warming scenario. The approximation used here is valid only at pH values around 8, and for small perturbations to the observed level of ocean DIC, and is consistent with the ocean pH values at present or those that are anticipated in the coming decades. For this pH range we may assume

$$\left[HCO_3^-\right], \left[CO_3^{2-}\right] \gg \left[H^+\right], \left[OH^-\right], \left[H_2CO_3^*\right].$$

The ions on the left are measured in hundreds to thousands of micromoles per liter (Figure 5.3). The smallness of $\left[H_2CO_3^*\right]$ may be deduced from Figure 5.3 (see green line near pH $= 8$). That $\left[H^+\right]$ is small is clear from the pH level,

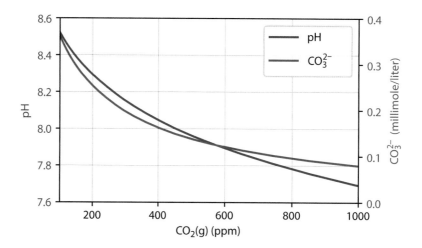

Figure 5.4: The response of pH and carbonate ion to CO$_2$ increase.
The solution of the carbonate system for a fixed alkalinity as in Figure 5.3, showing the ocean pH (blue) and the carbonate ion CO$_3^{2-}$ concentration (red) as a function of atmospheric CO$_2$.

which implies that $\left[\text{H}^+\right] \sim 10^{-8}$ mole/l. That $\left[\text{OH}^-\right]$ is small is similarly deduced from the water dissociation equation (5.16) and the fact that $K_w \sim 2 \times 10^{-14} \ (\text{mol/l})^2$.

With this approximation, let the carbonate system unknowns be the values of the five concentrations $\left[\text{CO}_2(\text{g})\right]$, $\left[\text{H}_2\text{CO}_3^*\right]$, $\left[\text{HCO}_3^-\right]$, $\left[\text{CO}_3^{2-}\right]$, and $\left[\text{H}^+\right]$. With a total of five unknowns ($\left[\text{OH}^-\right]$ is not calculated, nor needed now), we need five equations (the one for water dissociation is not needed),

$$K_H = \frac{\left[\text{H}_2\text{CO}_3^*\right]}{\left[\text{CO}_2(\text{g})\right]},$$

$$K_1 = \frac{\left[\text{HCO}_3^-\right]\left[\text{H}^+\right]}{\left[\text{H}_2\text{CO}_3^*\right]},$$

$$K_2 = \frac{\left[\text{CO}_3^{2-}\right]\left[\text{H}^+\right]}{\left[\text{HCO}_3^-\right]},$$

$$Alk_C = \left[\text{HCO}_3^-\right] + 2\left[\text{CO}_3^{2-}\right],$$

$$C_T = \left[\text{HCO}_3^-\right] + \left[\text{CO}_3^{2-}\right]. \tag{5.19}$$

The last two equations give

$$[\text{HCO}_3^-] = 2C_T - Alk_C, \tag{5.20}$$

$$[\text{CO}_3^{2-}] = Alk_C - C_T. \tag{5.21}$$

Using the K_2 equation,

$$[\text{H}^+] = K_2 \frac{2C_T - Alk_C}{Alk_C - C_T}. \tag{5.22}$$

Next, using the K_1 equation,

$$[\text{H}_2\text{CO}_3^*] = \frac{K_2}{K_1} \frac{(2C_T - Alk_C)^2}{Alk_C - C_T},$$

which, using Henry's law, gives

$$[\text{CO}_2(g)] = \frac{K_2}{K_1 K_H} \frac{(2C_T - Alk_C)^2}{Alk_C - C_T}. \tag{5.23}$$

We have now solved for all unknowns in terms of the equilibrium constants and the specified total CO_2 (DIC) and alkalinity. Given their definitions, total CO_2 (DIC) is necessarily smaller than alkalinity in this approximation (see last two lines in eqn 5.19), $Alk_C > C_T$. At the same time, the typical values given below indicate that $Alk_C < 2C_T$. The last equation (5.23) therefore makes it clear, for example, that if the DIC increases, the atmospheric CO_2 increases as well; we explore more such responses of the carbonate system to various perturbations in the next section.

   To calculate the numerical values of the above solution, one may use the typical values $Alk = 2350\ \mu\text{mol/l}$, $C_T = 2075\ \mu\text{mol/l}$, and the constants that are derived for a temperature and salinity of $T = 15\,^\circ\text{C}$ and $S = 35$ ppt, at a depth of $0\,\text{m}$: $K_H = 0.0375\ \text{mol/l/ppt}$, $K_1 = 1.15 \cdot 10^{-6}\ \text{mol/l}$, $K_2 = 7.43 \cdot 10^{-10}\ \text{mol/l}$, $K_w = 2.37 \cdot 10^{-14}\ (\text{mol/l})^2$; the calcite and aragonite solubility constants are $K_{sp,c} = 4.31 \cdot 10^{-7}\ (\text{mol/l})^2$ and $K_{sp,a} = 6.72 \cdot 10^{-7}\ (\text{mol/l})^2$, correspondingly.

5.3 RESPONSE TO PERTURBATIONS

We now consider the response of the ocean carbonate system to various perturbations that may be expected as part of anthropogenic climate change, including increased atmospheric CO_2 concentration, warming, and the long-term recovery from increased CO_2.

5.3.1 Response to increased atmospheric CO_2 concentration

If the atmospheric CO_2 increases, so would the ocean reservoir of total CO_2 due to Henry's law. Consider therefore that the DIC increases due to anthropogenic emissions by one unit. The alkalinity does not change in this scenario, as no ions related to weak acids are added to the ocean. The approximate solution of section 5.2.2 then allows us to calculate the response of the pH and carbonate ion concentration,

$$\Delta C_T = \mathbf{1} \uparrow, \quad \Delta Alk_C = \mathbf{0}$$

$$\left[H^+\right] = K_2 \frac{2C_T - Alk_C + \mathbf{2}}{Alk_C - C_T - \mathbf{1}} \uparrow \Rightarrow pH \downarrow$$

$$\left[CO_3^{2-}\right] = Alk_C - C_T - \mathbf{1} \downarrow.$$

The solution thus indicates that the pH should drop, and the ocean should become more acidic, as expected. The decrease in the concentration of the carbonate ion CO_3^{2-} means that the reaction of dissolution of calcium carbonate,

$$CaCO_3 \rightleftharpoons Ca^{2+} + CO_3^{2-}, \tag{5.24}$$

will be driven toward its right-hand side and therefore lead to more dissolution (or less deposition) of calcium carbonate, moving upward the transition zone between the dissolving deep ocean and the upper ocean. This is also seen by writing this reaction as $K_{sp} = \left[Ca^{2+}\right]\left[CO_3^{2-}\right]$, where reduction in the carbonate ion would lead to calcium carbonate dissolution, which increases the concentration of both ions and restores the product to its equilibrium value between the solid phase of calcium carbonate and its ions.

The fact that much of the dissolved CO_2 is converted into bicarbonate and carbonate ions via the carbonate system reactions means that for a given change to the atmospheric CO_2, the ocean absorbs much more CO_2 than it would have without these reactions, an effect referred to as the *buffer capacity* of the ocean carbonate system. Henry's law (eqn 5.4) controls the concentration of $\left[H_2CO_3^*\right]$, and without the carbonate reactions, the increase in dissolved CO_2 would correspond to the increase in $\left[H_2CO_3^*\right]$ alone. However, as noted above, the distribution of carbon among $\left[H_2CO_3^*\right]$, $\left[CO_3^{2-}\right]$, and $\left[HCO_3^-\right]$ is approximately 1%, 10%, and 90%, correspondingly (Figure 5.3). A small increase in $H_2CO_3^*$ due to the increase in atmospheric CO_2 and Henry's law is therefore accompanied by a larger increase in bicarbonate ions, and thus in a larger increase in dissolved CO_2 than would have been deduced by the change in $H_2CO_3^*$ alone. The bottom line is that most of the newly dissolved atmospheric CO_2 is converted into bicarbonate ions. The actual increase in carbon in the ocean carbonate system is therefore much larger than it would have been due to Henry's law and without these reactions, allowing the ocean to absorb more of the emitted anthropogenic CO_2.

5.3.2 Response to warming

So far the ocean has been absorbing a significant fraction of the anthropogenic CO_2 emission, leading to the observed acidification yet also to a reduction in the greenhouse effect and warming that would have been experienced otherwise. However, a warming of the ocean would lead to changes in the solubility of CO_2 in seawater. As the warming intensifies, less emitted CO_2 will be dissolved, and some dissolved CO_2 may be released to the atmosphere, further amplifying the greenhouse warming. During the last glacial maximum 21,000 yr ago, for example (Box 10.1), the ocean temperature was colder by a few degrees, and the CO_2 concentration was 180 ppm, about 100 ppm less than its preindustrial value of 280 ppm. About a third of this drop in CO_2 can be attributed to the cooler glacial ocean temperatures, an effect which is referred to as the *solubility pump*.

The effect of the dependence of solubility on temperature is demonstrated in Figure 5.5, where the pH, atmospheric CO_2, and reaction constants are shown as a function of temperature, in a calculation specifying constant DIC and alkalinity. The dependence of the reaction constants on temperature leads to

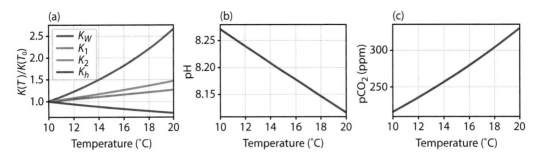

Figure 5.5: Response of the carbonate system to warming, showing quantities as a function of the ocean temperature, when the DIC and alkalinity are assumed fixed. (a) Reaction constants normalized by their values at $10\,^\circ$C. (b) pH. (c) Atmospheric partial pressure of CO_2, denoted pCO_2.

the changes seen to ocean pH and to the atmospheric CO_2 with temperature, showing increasing atmospheric CO_2 with warming, as expected. The figure suggests that a few degrees of warming can lead to a significant atmospheric CO_2 increase of a few tens of parts per million, even without the direct contribution of anthropogenic CO_2 emissions.

The approximate solution to the carbonate system discussed in section 5.2.2 reproduces the magnitude of the changes to the pH and pCO_2, with temperature seen in the exact solution shown in Figure 5.5 (albeit with a non-negligible constant bias, not shown), and can provide further insight into this effect. Equation (5.22) suggests that the pH changes as a function of temperature mostly due to the variation of K_2 with temperature, while equation (5.23) indicates that K_1, K_2, and K_H all play a role in leading to the change in atmospheric CO_2 with temperature, rather than, say, only Henry's law constant being responsible for the direct dissolution of CO_2 in seawater, as one might think naively. Note that in the fraction appearing in the solution for the atmospheric CO_2 concentration, K_2/K_1K_H, Henry's constant K_H decreases with temperature, while the other two increase. The ratio overall increases, leading to the increase in atmospheric CO_2 with warming.

5.3.3 Long-term decline of anthropogenic CO_2

Human release of carbon into the atmosphere so far amounts to an order of 700 GtC (the atomic weight of carbon is 12 atomic mass units, while that of carbon dioxide is 44; implying a 1 ton carbon in $44/12 = 3.67$ tons of CO_2).

The current anthropogenic emission rate is about 10 GtC/yr, and business-as-usual scenarios anticipate over 1000 GtC total further emission by year 2100. These emissions far exceed the emission of CO_2 by the main natural source due to volcanoes (Box 5.1). The lifetime of emitted CO_2 affects long-term warming projections and is clearly of interest. The first stage in the adjustment to the added anthropogenic carbon is a relatively fast (years) absorption by the terrestrial biosphere and an equilibration with the upper ocean where some of the CO_2 dissolves, leading to the lowering of ocean pH, as we have seen above.

The deeper ocean takes up and responds to the dissolved CO_2 more slowly (hundreds of years) due to the slow mixing between the upper and deeper ocean (section 3.1.2), until a final partitioning of CO_2 is achieved between the atmosphere and ocean. The lowering of the ocean pH leads to a dissolution of calcium carbonate in ocean bottom sediments (or reduced burial of calcium carbonate), as discussed in section 5.3.1. This dissolution restores the ocean pH, which may be understood using our simplified carbonate system as follows. The dissolution of one unit of $CaCO_3$ into Ca^{2+} and CO_3^{2-} implies the addition of one unit of total CO_2 (ΔC_T, in the form of the carbonate ion) and two units of carbonate alkalinity (ΔAlk_C, due to the double negative charge of the carbonate ion). The resulting pH and atmospheric CO_2, based on our simplified solution, are then

$$\Delta C_T = \mathbf{1} \uparrow, \quad \Delta Alk_C = \mathbf{2} \uparrow$$

$$\left[H^+ \right] = K_2 \frac{2C_T - Alk_C + \mathbf{0}}{Alk_C - C_T + \mathbf{1}} \downarrow \Rightarrow pH \uparrow$$

$$\left[CO_2(g) \right] = \frac{K_2}{K_1 K_H} \frac{(2C_T - Alk_C + \mathbf{0})^2}{Alk_C - C_T + \mathbf{1}} \downarrow .$$

That is, pH increases (because $\left[H^+ \right]$ decreases) in response to calcium carbonate dissolution, and the atmospheric CO_2 (eqn 5.23) decreases. The timescale for this process, which involves interaction with ocean sediments, though, is thousands of years, and more detailed calculations show that a significant fraction of emitted anthropogenic CO_2 may therefore remain in the atmosphere for tens of thousands of years. Eventually, on an even longer timescale, the natural sink of CO_2, involving its dissolution in rain water, formation of a weak acid that interacts with silicate rocks, leads to the formation of calcium carbonate whose dissolved components are washed into the ocean, leading to increased $CaCO_3$ burial there and restoring the atmospheric CO_2 to its natural level. This natural level has very slowly varied over millions of years due, for example, to changes in emission rates by volcanoes or other tectonic processes, leading to higher CO_2 during past warm climate periods such as the Pliocene (5.3–2.6 Myr) and Eocene (56–33.9 Myr), discussed in Box 4.1.

Final word

In the case of ocean acidification, unlike many other issues discussed in this book, the uncertainty in the response to anthropogenic forcing is small: given an atmospheric CO_2 increase, the pH will drop. The signal of ACC is already clearly seen as the reduction in the observed ocean surface pH (blue markers in Figure 5.1); given that it is not difficult to calculate the expected acidity change due to the increase in atmospheric CO_2, the attribution to ACC (specifically to the increase in atmospheric CO_2) in this case is straightforward. As mentioned, the response of the ocean biology to the change in ocean acidity is a more complex issue, yet if CO_2 increases significantly following some high-end emission scenario, the ocean pH drop will be large, and one may anticipate a non-negligible disruption to ocean biology.

5.4 WORKSHOP

A Jupyter notebook with the workshop and corresponding data file are available; see https://press.princeton.edu/global-warming-science.

1. **Characterizing pH changes:**

   (a) Plot a time series of the globally averaged surface ocean pH from 1850 to 2100, combining observations and RCP8.5 projection.

   (b) Calculate the percent change in $[H^+]$ concentration involved in the pH change observed so far from 8.16 to 8.06.

   (c) Plot maps of surface ocean pH for 1850 and 2000 (observed) and for the RCP8.5 projection for 2100.

   (d) Discuss the amplitude of current spatial variations versus expected temporal changes in the context of the expected robustness of ocean biology to acidification.

   (e) Calculate and contour the percentage change in $[H^+]$ concentration due to pH changes from year 1850 to 2000 and to 2100.

2. **The carbonate system solution:**

   (a) Plot the given exact solution for the carbonate species CO_3^{2-}, HCO_3^-, and $H_2CO_3^*$ and the DIC, as a function of pH, for the case of fixed *Alk* and varying DIC.

   (b) Plot separately the pH and CO_3^{2-} as a function of atmospheric CO_2 from the same solution.

3. **The carbonate system buffer effect:** Plot the expected change to the total dissolved CO_2 (that is, to the DIC) as a function of the atmospheric CO_2 concentration in the range 200 to 1000 ppm, based on the given exact solution. Superimpose the expected change calculated based on Henry's law alone (represented in this case by the change to $H_2CO_3^*$ only, as other species do not change by assumption) that would have occurred without the carbonate system reactions.

4. **Understanding the response:** Calculate and plot the approximate solution to the carbonate system:

   (a) Explain the conditions needed for the approximate solution discussed in section 5.2.2 to be valid, and identify the range of atmospheric CO_2 concentrations for which you expect these conditions to hold.

(b) Use the approximate solution of the carbonate system to calculate the concentration of the carbonate species as a function of pH for $Alk = 2.3$ mM/l, in the range of atmospheric $CO_2(g)$ in which you expect the approximate solution to be valid. To do so, vary the DIC over an appropriate range of values, and calculate both the carbonate species and pH. Superimpose dashed lines representing the provided exact solution.

(c) Plot the approximate solution for the pH as a function of atmospheric $CO_2(g)$. Superimpose a plot of the exact solution.

5. **Estimating observed and future pH from CO_2:**

(a) Use the approximate solution with $Alk = 2.3$ mM/l, varying the DIC (C_T) to calculate the pH and $CO_2(g)$. Find the C_T values that result in the observed CO_2 concentrations at years 1850 and 2000 and the RCP8.5 value at 2100. Deduce the pH values at these years. Compare to the above ocean pH time series for 1850–2100.

(b) *Optional extra credit:* Use a similar procedure to the approximate solution of the carbonate system equation in section 5.2.2 to solve for the pH given the Alk and $CO_2(g)$ (instead of using given Alk and C_T as in that section). The solution involves deriving a quadratic equation for $[H^+]$ whose coefficients include Alk and $CO_2(g)$ and the various reaction coefficients. Let $Alk = 2.3$ mM/l and plot the expected upper ocean pH given the observed CO_2 concentration values from 1850 to present and the RCP8.5 values from present to 2100. Compare to the given pH values for these periods obtained from measurements and more sophisticated modeling by superimposing a plot of the given surface ocean pH for 1850–2100.

6. **Guiding questions to be addressed in your report:**

(a) What is ocean acidification, and why is it an important concern?

(b) What is the current state of ocean acidification relative to 1850, and what is expected by year 2100 in a business-as-usual scenario?

(c) What is the ocean carbonate system?

(d) How and why does the concentration of the carbonate ion CO_3^{2-} affect the dissolution of calcium carbonate?

(e) Based on the approximate solution to the carbonate system, how does the addition of CO_2 to the atmosphere change the dissolved

CO_2, the bicarbonate ion, the carbonate ion, and thus calcium carbonate dissolution? Discuss the buffer effect.

(f) What are the economic consequences of ocean acidification? (Research this beyond what was presented in class.)

(g) What are the implications of ocean acidification for the utility of solar geoengineering?

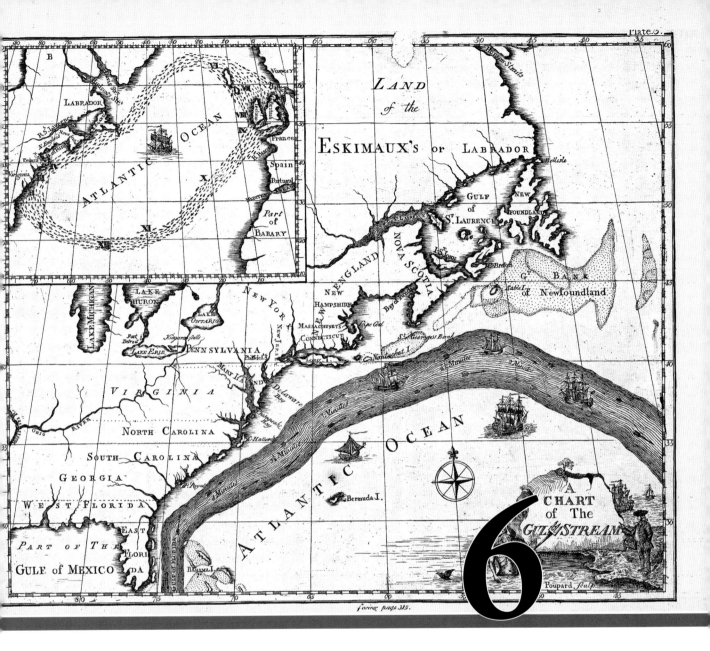

OCEAN CIRCULATION

Key concepts

- The Atlantic meridional overturning circulation
- Observations, has the Atlantic meridional circulation started collapsing? Future projections
- Ocean temperature, salinity, density
- Tipping points, multiple equilibria, stability, hysteresis
- Consequences of meridional circulation collapse
- The oceans in global warming

The Atlantic meridional overturning circulation (AMOC), also referred to as the *thermohaline circulation* or the ocean conveyor belt, is the zonally integrated northward flow of water in the upper 1 km of the North Atlantic Ocean, which is cooled in the northern North Atlantic, becomes denser and sinks, and returns south between 2 and 3 km depth as the North Atlantic Deep Water (NADW; see Box 6.1). The circulation magnitude is about 20 Sverdrups (Sv), where 1 Sv is 10^6 cubic meters of seawater per second. For perspective, all the world rivers combined transport about 1 Sv, and the Gulf Stream transports up to about 100 Sv.

Because the AMOC involves the flow of warm water northward and a return of cold water southward, it carries a significant amount of heat poleward, of the order of 1 petawatt, or 10^{15} W. This heat is released to the atmosphere in the northern North Atlantic, where the water cools and sinks. AMOC is therefore often credited for contributing to the warmth of Northern Europe relative to areas of eastern North America that are at the same latitude or even farther south (London is at 51.5°N with average maximum/minimum January temperatures of 9 °C/4 °C, while in Boston, at 42°N, these temperatures are 3 °C/−5 °C). Climate projections suggest that this circulation might weaken dramatically by 2100, mostly due to the increase in fresh water input into the northern North Atlantic from increased precipitation and ice melting, which would make the surface ocean water fresher and therefore lighter and would not allow it to sink, weakening the northward ocean heat transport. The reason Northern Europe is warmer than eastern North America turns out to involve additional factors, including the atmospheric circulation and the location of the atmospheric jet stream, but as we will see, a collapse of AMOC may still lead to local cooling of sea surface temperatures in the northern North Atlantic Ocean.

Furthermore, the ocean meridional circulation may be an example of the climate system encountering a tipping point—a *gradual* change in CO_2 concentration past a certain threshold may lead to an *abrupt* AMOC collapse. Such an abrupt response may be accompanied by irreversibility of the change: if the CO_2 concentration stops increasing and is gradually *reduced* past the threshold where the abrupt change occurred, AMOC may not necessarily recover its original value. While it is far from obvious that AMOC will experience such abrupt changes and tipping point dynamics, we use it below as an opportunity to understand how climate tipping points are analyzed, why they occur, and what exactly happens when they do.

The sinking of deep water in the North Atlantic is supplemented by similar sinking to even greater depths near Antarctica (Box 6.1), and the Atlantic meridional overturning circulation is part of a global meridional overturning ocean circulation (MOC). Some of the deep return flow from the North Atlantic upwells in the Southern Ocean, driven by the winds there, and some proceeds to the Pacific and Indian Oceans, gradually warms and slowly upwells over large areas on its way there, and returns as an upper ocean warm flow to the North Atlantic Ocean. This overturning circulation co-exists with the wind-driven ocean circulation in the great upper ocean (top ~1 km) gyres. In the western North Atlantic, for example, the narrow (~100 km width) Gulf Stream transports up to about 100 Sv northward, compensated by a slow southward return flow over the upper North Atlantic Ocean east of the Gulf Stream. Similar wind-driven upper ocean gyres exist in all ocean basins, with intense narrow flows on the west and a slow broad return flow in the basin interior. Had we followed a specific water parcel in the ocean, we might have seen it circling a few times over a few decades in these horizontal wind-driven gyres, then escaping northward and sinking to the deep ocean, eventually returning hundreds of years later to the surface in a different ocean basin.

In the following section we discuss present observations and future warm-climate projections of AMOC (section 6.1). Next, we analyze a highly idealized yet surprisingly insightful model of the AMOC, allowing us to understand how tipping points occur in this case and more generally in the climate system (sections 6.2 and 6.3), and discuss the role of simple climate models (6.4). We conclude with a brief discussion of the consequences of an AMOC collapse in full-complexity climate models (section 6.5) and with a survey of the roles the ocean plays in anthropogenic climate change (section 6.6).

Climate Background Box 6.1
Ocean temperature, salinity, and water masses

The zonally averaged temperature and salinity across the Atlantic Ocean are contoured in the accompanying figure as a function of latitude and depth in panels a and c, while the profiles of temperature and salinity versus depth in panels b and d are averaged within the Atlantic over latitudes $30°S$–$0°$. The deep ocean is seen to be much colder than the surface, being filled via sinking at the cold high latitudes. The temperature away from the high latitudes is typically vertically uniform in the upper 25–50 m *mixed layer* that is well mixed by wind and waves. It then decreases rapidly in the upper kilometer, in a depth range known as the *main thermocline*. Below the thermocline, the temperature is nearly uniform, gradually decreasing toward the bottom. Salinity (measured in parts per thousands; say a kg of salt in a metric ton of seawater) variations are a result of surface evaporation, which increases the salinity (and the freezing of sea ice that results in the rejection of brine), and precipitation and river runoff, which decrease it. Water bodies with characteristic temperature and salinity are referred to as *water masses*. A tongue of salty water spreading south from the North Atlantic is seen at a depth of 2–3 km, known as the North Atlantic Deep Water (NADW). Two fresher water masses are seen to spread northward from the south, the Antarctic Intermediate Water (AAIW), at a depth of about 1 km and above the NADW, and the Antarctic Bottom Water (AABW), near the bottom and under the NADW. Note that the vertical dimension is stretched significantly, as the actual aspect ratio between the length and depth of the Atlantic Ocean is about 1000:1. These water masses are therefore essentially thin sheets of ocean water spreading horizontally with very little mixing due to small-scale turbulence across density surfaces. The spreading largely occurs over surfaces of constant density, such as the three marked by the black contours. The Stommel box model discussed in section 6.2 idealizes the formation and spreading of the NADW. A shutdown of the Atlantic meridional overturning circulation may disrupt this very distinct large-scale water mass structure.

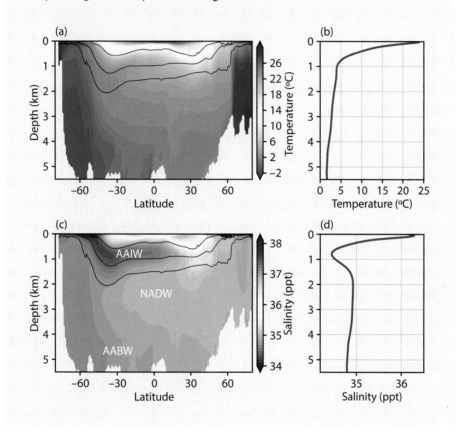

6.1 OBSERVATIONS AND PROJECTIONS

The AMOC may be defined as the total northward water volume transport in the upper 1 km of the North Atlantic Ocean, often calculated at one specific latitude. Since 2004, AMOC has been estimated continuously in the RAPID program using observations across 26.5°N in the North Atlantic Ocean, including regular ship surveys that measure temperature, salinity, and density across the Atlantic Ocean and observations of the wind that drives upper ocean transports, of

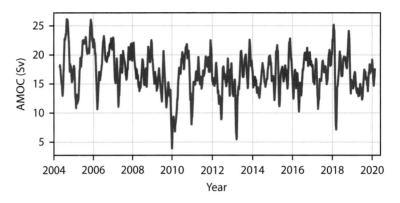

Figure 6.1: The observed AMOC.
A time series from the RAPID campaign.

the Gulf Stream, and more. The observed AMOC time series shows a strong, yet noisy and irregular, seasonal cycle (Figure 6.1). Historical ship observations have been used to estimate the transport across 25°N over the past few decades—for 1957, 1981, 1992, 1998, and 2004—suggesting a detection of a significant AMOC slowdown from 23 Sv to 15 Sv over this period. However, this seeming weakening is likely a result of the large-amplitude seasonal cycle we are now aware of thanks to the RAPID campaign (see this chapter's workshop), and for now it is probably fair to say that we are unable to detect an AMOC slowdown with confidence, even if future projections of a slowdown seem robust across climate models. While the observed record is still short, AMOC oscillations seen in climate models have periods from a couple of decades to centuries. These oscillations affect the large-scale sea surface temperature in the North Atlantic in particular. A variability mode known as the *Atlantic multidecadal oscillation* (AMO) is a pattern of alternating warming and cooling with an amplitude of some 0.2 °C that is inferred from sea surface reconstructions over the past century or so. This mode was suggested to be related to AMOC variability, although its very existence and driving mechanisms are being debated. While speculative, the AMO was suggested to affect the occurrence of droughts and hurricanes, among other influences.

Future RCP scenarios (e.g., RCP8.5) suggest a dramatic future slowdown of the AMOC (Figure 6.2), mostly due to a freshening of the northern North Atlantic by increased precipitation and ice melting. The freshening reduces the

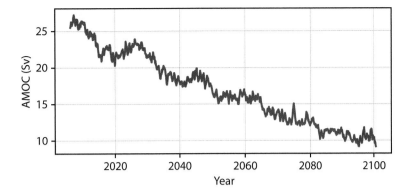

Figure 6.2: An AMOC future projection.
A projected AMOC transport time series in a climate model run under the RCP8.5 scenario.

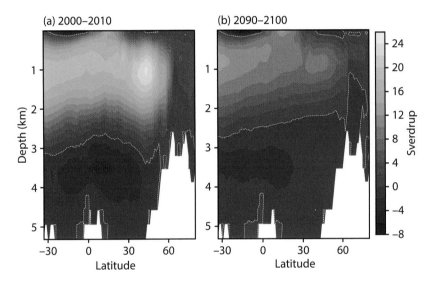

Figure 6.3: The AMOC streamfunction.
Shown in units of Sv from a climate model run under the RCP8.5 scenario, during the first and last decades of the 21st century.

density of the surface water in the northern North Atlantic, and this leads to a lowering of the north-south density difference and makes the high-latitude surface water lighter, preventing its sinking.

AMOC is often represented in model simulations and projections via a *streamfunction*, which is depicted by a contour plot as a function of latitude and depth across the North Atlantic Ocean (Figure 6.3). The zonally integrated AMOC transport occurs along contours of constant streamfunction values, and the difference between contour values (say at 24°N) at two depths is equal to the total meridional mass transport in this depth range. The AMOC streamfunction is shown in Figure 6.3 for the first and last decades of the 21st century, for an RCP8.5 projection, demonstrating the projected weakening.

AMOC is driven by a complex set of factors, including north-south ocean water density differences due to temperature and salinity differences between the low and high latitudes in the North Atlantic Ocean (Box 6.1), winds in the Southern Ocean, and small-scale ocean mixing due to turbulence on a scale of meters or less within the ocean interior. Such ocean mixing is therefore another sub-grid scale process that has global climate implications and needs to be parameterized in ocean models. As an introduction to the analysis in the next section, we note that a somewhat naive, yet perhaps not useless, view of the AMOC is that it is dominantly driven by a north-south ocean water density difference, which leads to a pressure gradient that can drive a northward flow. In this view, the cold polar water versus warm tropical oceans leads to a meridional density gradient that *drives* AMOC. On the other hand, the salty tropical ocean versus the relatively fresh high-latitude ocean has the opposite effect on the meridional density gradient, *braking* the AMOC.

The presence of the Earth rotation and the resulting Coriolis force seem to suggest that this view is too naive (a north-south pressure gradient would drive a zonal ocean current; see section 4.2.1). Yet there are reasons to believe that the north-south density gradient still plays a dominant role, and relying on it still leads to very robust insights.

6.2 THE STOMMEL MODEL

Our objective is to understand what might happen to the AMOC in a warmer climate, when a freshening of the northern North Atlantic is expected due to ice melting and increased precipitation. In particular, we are interested in the prospect of an abrupt collapse of AMOC as CO_2 gradually increases beyond a certain threshold, a scenario often referred to as the crossing of a tipping point.

Divide the North Atlantic Ocean very crudely into two boxes (Figure 6.4), one representing the high latitudes (box 1) and the other the mid-latitudes/subtropics (box 2). The boxes are assumed to extend from the ocean surface to

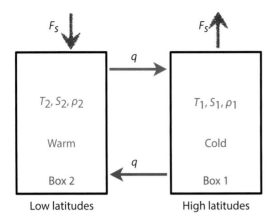

Figure 6.4: Stommel box model schematic.

the ocean bottom and be well mixed (clearly a crude approximation, as we know the ocean temperature, salinity, and density to vary significantly with depth; see Box 6.1). The two boxes are connected via a circulation representing AMOC, denoted by the red and blue arrows, which transports a volume flux q between the boxes. The AMOC volume transport $(\mathrm{m}^3/\mathrm{s})$ is assumed proportional to the density difference between the boxes, $q = K(\rho_1 - \rho_2)$. The density of seawater is approximated to be a linear function of temperature and salinity, $\rho(T, S) = \rho_0 - \alpha_T(T - T_0) + \beta_S(S - S_0)$, increasing with salinity and decreasing with temperature.

Surface forcing: salinity and evaporation/precipitation

To see how evaporation and precipitation affect salinity, consider a bucket filled with a volume V of seawater and undergoing evaporation. First, the volume budget of the water in the bucket is

$$\frac{dV}{dt} = -EA,$$

where E is the net evaporation rate per unit area and A is the surface area of bucket. Letting the mass of salt per unit mass of seawater be denoted by the salinity S, the total salt in the bucket, not affected by evaporation, is equal to ρSV, where ρ is the water density. Assuming the density is nearly constant, the salt conservation statement is, therefore,

$$\frac{d}{dt}(SV) = 0,$$

which we can expand using the above mass conservation equation into

$$V\frac{dS}{dt} = -S\frac{dV}{dt} = SEA \approx S_0EA.$$

Salinity in the ocean varies mostly between 34 and 36 ppt (1 ppt roughly representing a kg salt per metric ton of seawater), and $S_0 = 35$ ppt denotes an averaged reference salinity. We term S_0E the *virtual salt flux* per unit area. Because evaporation rates are not large, the changes to the ocean salinity are typically relatively small and the above approximation of using S_0 instead of S is a very good one, accurately describing ocean salinity changes due to evaporation and precipitation. A typical order of magnitude of evaporation in the ocean is 1 m/yr, and given the ocean depth of 4000 m, evaporation does not significantly affect ocean volume. This also means that the water volume exchanges involved in evaporation and precipitation may be neglected, and we may represent their effect by the virtual salt flux, which is denoted F_s in Figure 6.4 and in the derivation below, also referred to as the *fresh water forcing*.

Salt budget equations for the Stommel model

We assume that the ocean temperature is set by the interaction with the atmosphere, which is cold in polar areas and warmer in subtropical areas. We therefore let the ocean box temperatures T_1 and T_2 be fixed to a first approximation, so that circulation changes occur only due to ocean salinity changes. The only unknowns, therefore, are the salinities of the two boxes and the circulation q. Write the salt budget equation for box 1, first assuming that $q > 0$, representing a salt transport qS_2 into box 1 and a transport qS_1 from box 1 to box 2. A freshening effect induced by excess precipitation in the high-latitude box 1 is represented by a $(-F_s)$ virtual salt flux contribution to the budget,

$$V_1\frac{dS_1}{dt} = qS_2 - qS_1 - F_s.$$

If $q < 0$, we have salt transport of $(-q)S_2$ into box 1 and a transport $(-q)S_1$ out of box 1 to box 2,

$$V_1\frac{dS_1}{dt} = (-q)S_2 - (-q)S_1 - F_s.$$

Together, this may be combined into a single form, which we write for both boxes as

$$V_1 \frac{dS_1}{dt} = |q|(S_2 - S_1) - F_s$$

$$V_2 \frac{dS_2}{dt} = |q|(S_1 - S_2) + F_s. \tag{6.1}$$

Note the absolute value of the transport appearing here, combining the above two cases of positive and negative AMOC volume transport q. The fresh water forcing is assumed positive, $F_s > 0$, corresponding to a freshening of the polar box and to making the low-latitude box saltier.

Solution

Take the difference of the two salt budget equations (6.1), define $\Delta T = T_1 - T_2 < 0$ and $\Delta S = S_1 - S_2$, and let the box volumes be the same, $V_1 = V_2 \equiv V$, for simplicity. The transport may then be written as

$$q = K(\rho_1 - \rho_2) = K(-\alpha_T \Delta T + \beta_S \Delta S),$$

and the equation for the salinity difference becomes

$$-V \frac{d\Delta S}{dt} = 2|q|\Delta S + 2F_s$$

or

$$-V \frac{d\Delta S}{dt} = 2K|(-\alpha_T \Delta T + \beta_S \Delta S)|\Delta S + 2F_s. \tag{6.2}$$

Define a rescaled known temperature difference variable, $X = \alpha_T \Delta T < 0$, and an unknown rescaled salinity difference, $Y = \beta_S \Delta S$. In a steady state, $d\Delta S/dt = 0$, so that equation (6.2) leads to a simple quadratic equation for $Y = \beta_S \Delta S$,

$$|Y - X|Y = -\frac{\beta_S F_s}{K}.$$

Because of the absolute value, there are two cases to consider, $q > 0$ and $q < 0$, equivalent to $Y > X$ and $Y < X$. Noting again that $X < 0$, we first consider the case $Y < X$,

$$Y^2 - XY - \frac{\beta_S F_s}{K} = 0$$

$$Y = \frac{X}{2} - \frac{1}{2}\left(X^2 + 4\frac{\beta_S F_s}{K}\right)^{1/2}. \tag{6.3}$$

Next, consider $Y > X$,

$$Y^2 - XY + \frac{\beta_S F_s}{K} = 0$$

$$Y = \frac{X}{2} \pm \frac{1}{2}\left(X^2 - 4\frac{\beta_S F_s}{K}\right)^{1/2}. \tag{6.4}$$

Note that the plus solution for Y in the first case is positive and is therefore not consistent with the assumption $Y < X$ used to obtain that solution, so there are no more than three solutions for a given value of the fresh water forcing F_s.

6.3 MULTIPLE EQUILIBRIA, TIPPING POINTS, HYSTERESIS

Multiple equilibria

Given the solution for $Y = \beta_S \Delta S$ as a function of the fresh water forcing F_s (eqns 6.3 and 6.4, plotted in Figure 6.5a), we can calculate the salinity difference ΔS, and from that the circulation q as a function of the fresh water forcing F_s, as in Figure 6.5b. The number of real-valued solutions varies from one to three, depending on the value of the fresh water forcing. Note that the solution shown by the green line in Figure 6.5b represents a northward upper ocean flow, while that shown in red is a weak, reversed flow that is the only possible solution for large fresh water forcing. The results are already remarkable at this point: a given fresh water forcing can lead to three different solutions for the overturning circulation. As we will see, the existence of such *multiple equilibria* is the first ingredient needed for tipping points to occur.

Stability

The intermediate solution (blue in Figure 6.5a,b,c) is unstable. That is, a small perturbation from that steady state would result in the circulation transitioning to one of the other two solutions that exist for the same fresh water forcing. To

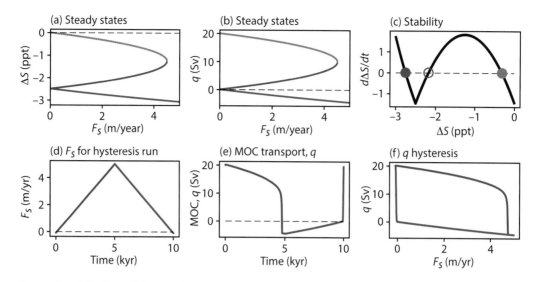

Figure 6.5: Solution of the two-box model.
(a, b) Steady states of salinity difference and MOC as a function of fresh water forcing F_s. (c) Stability analysis: $d\Delta S/dt$ as a function of ΔS for F_s corresponding to 2 m/s (to convert this to a virtual salt flux, multiply by the box area and by the reference salinity S_0), from eqn (6.2). (d) Time-dependent fresh water forcing used for the hysteresis run. (e, f) Results of the hysteresis run.

see this, consider $d\Delta S/dt$ plotted as a function of ΔS in Figure 6.5c using equation (6.2). If the solution is exactly at the steady state denoted by the empty blue circle, $d\Delta S/dt = 0$ and the steady state is maintained. However, suppose that the solution deviates a bit to the right (the salinity difference between the two boxes, ΔS, increases due to some random weather event affecting precipitation and evaporation). At that point, Figure 6.5c shows that $d\Delta S/dt > 0$, and the salinity difference will therefore keep increasing and getting away from this steady state. Similarly, if the solution deviates to the left of the steady state, marked by the empty blue circle in panel c, $d\Delta S/dt$ is negative there, and therefore the salinity difference will keep decreasing away from the steady state. The solution marked by the blue circle in Figure 6.5c and by the blue lines in panels a and b is therefore referred to as *unstable*. If we start near one of the stable solutions (filled red and green circles in Figure 6.5c), a small increase in ΔS leads to $d\Delta S/dt < 0$, while a small decrease in ΔS leads to $d\Delta S/dt > 0$. Thus, in both cases the deviation decreases back toward these two steady states, and they are therefore referred to as *stable* steady states.

Tipping points and hysteresis

The existence of multiple equilibria for a given fresh water forcing value F_s leads to both the possibility of abrupt changes as the forcing F_s changes gradually and to the possibility of irreversible changes to the circulation in a greenhouse scenario. To see this, suppose the circulation is at a steady state corresponding to a point on the upper (green) solution for q in Figure 6.5b and that we gradually and very slowly increase the fresh water forcing such that the solution for the overturning circulation is always at equilibrium with the fresh water forcing. The circulation solution then moves to the right along the green line and thus weakens until the fresh water forcing amplitude is at the critical value of $F_c \approx$ 4.2 m/s, at which the system switches from three solutions to only one. That one remaining solution is the very weak (reversed) circulation on the red curve in Figure 6.5b, so the circulation must abruptly switch to that solution—that is, collapse!

The lower three panels of Figure 6.5 and the schematic in Figure 6.6 demonstrate this abrupt collapse scenario. Suppose that in a global warming scenario, a gradual increase of CO_2 leads to a gradual increase in precipitation/melting and thus in the fresh water forcing F_s as a function of time (blue curve in Figure 6.5d). Figure 6.5e shows the transport as a function of time for this forcing, showing (blue line) a gradual decrease and then a collapse just before year 5000. This abrupt collapse, corresponding to the switch between the two equilibria solutions, denotes the occurrence of a tipping point. The schematic Figure 6.6 complements the picture. Starting with the bright green point (A) representing the present-day climate, the fresh water forcing increases slowly due to a CO_2 increase, and the AMOC solution responds by moving to the right (black solid arrows). As the solution reaches point (B), the stable green solution ceases to exist and the AMOC has to transition to the only stable solution available for larger fresh water values, which is a collapsed circulation shown by point (C) on the red line, denoting the other stable solution.

If the fresh water forcing is now made gradually weaker in time (red curve in Figure 6.5d), say, because CO_2 values are finally gradually decreasing, the circulation strengthens again (red curve in Figure 6.5e, dashed arrows in Figure 6.6) gradually at first. Even as the fresh water forcing is reduced below the critical value F_c (point (C) in Figure 6.6), the circulation does not recover (does not

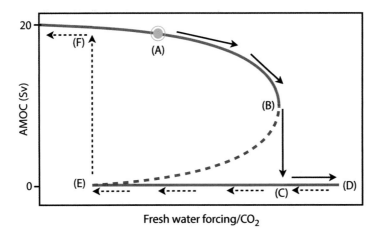

Figure 6.6: A schematic of the hysteresis and tipping point behavior in the two-box Stommel model.

jump back to the green line). Recovery happens only when $F_s = 0$ (point (E) in Figure 6.6). The different evolution of the solution for increasing and decreasing forcing, expressed as the loop showing the AMOC transport as a function of the fresh water forcing in Figures 6.5f and 6.6, is termed *hysteresis*. The existence of multiple stable and unstable solutions and the resulting hysteresis are all a result of the nonlinear nature of equation (6.2) for the salinity difference between the two boxes (that is, the right-hand side is quadratic in ΔS). In general, tipping points, hysteresis, and multiple equilibria are due to nonlinearities, of which there are numerous in the climate system.

6.4 KEEPING IT SIMPLE

Remarkably, some full-complexity ocean models, of the same complexity class used in climate studies, show abrupt changes to AMOC and a corresponding hysteresis loop when an appropriate scenario of increasing and then decreasing fresh water forcing is applied. The main difference from our above Stommel box model analysis is that AMOC seems to weaken and possibly vanishes at high fresh water forcing values in more realistic climate models rather than reverse as in the above box model. Furthermore, it was suggested that the present-day circulation may be close to the threshold that leads to such an abrupt irreversible collapse. On the other hand, the decline of AMOC in the full-complexity

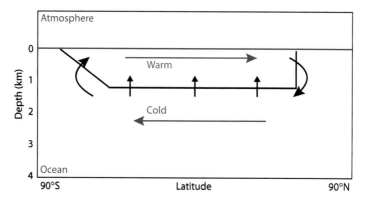

Figure 6.7: An alternative AMOC idealization.
A schematic of an idealized AMOC model that is more appropriate as an idealization of the ocean temperature structure than the Stommel model, showing a warm pool of water in the upper kilometer or so, overlying a deep cold ocean that outcrops in the high latitudes of both hemispheres. Meridional heat transport by the Atlantic Ocean is the result of the upper layer's northward flow of warm water and the deeper return flow of colder water (crudely representing the NADW). Some of the cold water returns to the surface in the Southern Ocean and some as upwelling along the way.

coupled ocean-atmosphere climate model, shown in Figure 6.2, is gradual rather than an abrupt collapse. It is possible that the CO_2 change in this scenario was too fast to allow the tipping point to be clearly expressed or that this model does not have a tipping point for AMOC, demonstrating the uncertainty in this prediction.

The Stommel model in Figure 6.4 is by no means a reasonable representation of the actual ocean thermal structure in panel a of the figure in Box 6.1. A more appropriate idealization is of a layer of warm water overlying a colder deep ocean, as in the schematic Figure 6.7; many other idealized models have been proposed for this problem as well. This brings up an important point about climate science: Observations and realistic climate model simulations are critical tools for us to be able to study and predict the present and future climates. But a deeper understanding is often achieved using idealized (toy) models that can be analyzed in depth. Of course, "everything should be made as simple as possible but not simpler" (attributed to Einstein), so is the Stommel model too simple? When an analysis similar to the above is applied to alternative idealized models, an AMOC decline at higher fresh water forcing (higher CO_2) is often seen, but it is not necessarily abrupt and there isn't always a hysteresis. As mentioned, it is also not clear that full-complexity coupled ocean-atmosphere climate models robustly show such a hysteresis of AMOC when forced with slowly increasing and decreasing CO_2, leaving the robustness of this prediction of an AMOC tipping point uncertain.

The existence of tipping points for increasing greenhouse forcing has been proposed in the context of other climate components as well, from Arctic sea ice to Arctic clouds to subtropical clouds to ice sheets, underlying the possibility that a gradual CO_2 change may lead to abrupt changes and thus surprises whose precise timing is difficult to predict. In nearly all cases the mathematical structure of the system allowing a tipping point behavior is similar to the schematic Figure 6.6, with the number of solutions changing as a function of the CO_2 from one to three (including an unstable one) to one. Whether a robust climate tipping point behavior is in fact expected in any specific climate component as CO_2 increases, with the accompanying irreversibility when CO_2 eventually decreases, is still an open issue.

6.5 CONSEQUENCES OF AMOC COLLAPSE

The sea surface temperature warming over the 21st century under the RCP8.5 scenario is shown in Figure 6.8. While there is a strong overall warming, there is also a large-scale patch in the North Atlantic south of Greenland where the SST has cooled. This local cooling is probably related to the ceased northward heat transport by the collapsed AMOC. (Importantly, the 2004 movie *The Day*

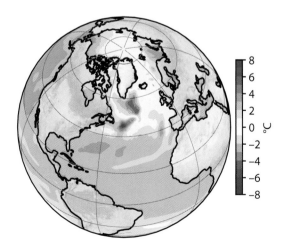

Figure 6.8: Consequences of AMOC collapse.
The SST at year 2100 minus that at 2006 in an RCP8.5 scenario.

After Tomorrow predicted that such a cooling due to AMOC collapse would lead to an ice age over North America within a few weeks. See Box 10.1 for a more reality-based picture of ice ages.)

While the above analysis of the Stommel box model assumed the temperature to be fixed, one does expect the SST to be affected by the collapse, as demonstrated by this image. The change to the temperature is not sufficiently large to invalidate our above analysis, yet it has a significant impact on the nearby ocean regions and continents. Other consequences involve the reduced ventilation of the deep ocean by the North Atlantic Deep Water currently sinking in the North Atlantic (Box 6.1), with possible effects on the large-scale deep ocean distributions of temperature, salinity, oxygen, and more. A similar-looking SST signal to the above projected one that has been observed in the North Atlantic in recent years is referred to as the North Atlantic warming hole. Whether it can be attributed to an already occurring AMOC weakening and/or to anthropogenic climate change is being debated.

6.6 THE OCEANS AND GLOBAL WARMING

It may be helpful to briefly survey some roles the world oceans play in the context of anthropogenic climate change, many of which are discussed in other chapters. We saw in section 3.1.2 that the deep ocean with its immense heat capacity can delay the atmospheric warming response to greenhouse forcing by taking a longer time to warm. Similarly, when atmospheric CO_2 will finally be decreasing, the already occurring ocean warming may delay the expected climate cooling. Ocean warming, of course, leads to the expansion of seawater and thus to sea level rise (chapter 4). By absorbing about a quarter of the anthropogenically emitted CO_2, the oceans reduce the greenhouse effect and therefore also delay warming, yet at the same time they experience ocean acidification, which causes its own set of problems (chapter 5).

Sea surface temperature plays a dominant role in several contexts, starting with hurricanes, whose maximum possible magnitude may be controlled by the temperature and heat capacity of the upper ocean, which fuel hurricanes via evaporative cooling and subsequent condensation and latent heat release in the atmosphere (section 8.2). Droughts are often forced by the deviation of SST, in areas that can be thousands of kilometers away from the drought location, from their long-term averaged values. Such deviation patterns of SST that lead to droughts are caused by natural climate variability modes, such as La Niña cooling in the eastern equatorial Pacific (Box 8.1), which forces droughts in California and elsewhere (section 12.2). Similarly, droughts and therefore forest fires in Australia can be forced by deviations of the sea surface temperature from its long-term mean in both the equatorial Pacific and Indian Oceans (section 14.3). If these natural SST variability modes changed in a warmer future climate, this would have a significant effect on regional precipitation, droughts, and fires.

Sea ice in the Arctic Ocean has been dramatically melting in past decades (chapter 9), possibly forced by anthropogenic climate change and putting in motion its own set of feedbacks that lead to local and remote effects on weather and climate. Sea ice around Antarctica has not responded similarly so far, demonstrating the complexity of the climate system and the uncertainty in attributing changes to anthropogenic climate change. The warming of ocean water near both Greenland and Antarctica can very effectively melt the massive (∼1 km thick) floating ice shelves there. This can weaken the force that these ice shelves exert on the ice flows that feed them from continental ice sheets. The weakening of this force, in turn, can lead to acceleration of ice flows from Antarctica and Greenland and to a further speedup of sea level rise (chapter 10).

Apart from this long yet still incomplete list of ocean effects, processes, and feedbacks involved in anthropogenic climate change, the oceans also add to the uncertainty involved in projecting future changes. The long timescales of the ocean response to climate forcing make it more difficult to model the eventual response of the climate system. Ocean eddies, the continuously changing ocean motions that are the equivalent of atmospheric weather systems, are characterized by much smaller length-scales (ocean eddies have scales of tens to hundreds of km, while weather systems extend over thousands of km). Even smaller-scale ocean motions also play a critical role, meaning that ocean/climate models must cover the global oceans and at the same time have a high resolution that can properly represent all relevant small-scale motions. This leads to the need to perform very computationally expensive and in fact infeasible simulations, forcing us to compromise on the representation of critical ocean motions and leading to less reliable projections. Our ability to model variability modes such as El Niño is getting better, yet we are still not in a position to reliably project future changes to these modes. This means that our ability to project future changes to droughts, regional precipitation, and fires is necessarily still limited. It is important to note that the reliability of future climate change projections is still mostly limited by our incomplete understanding of many issues rather than by computational resources. Ocean science therefore has many important, challenging, and exciting open problems requiring further progress.

6.7 WORKSHOP

A Jupyter notebook with the workshop and corresponding data file are available; see https://press.princeton.edu/global-warming-science.

Required "reading": *The Day After Tomorrow* 2004 movie :-)

1. **Observations:**

   (a) Estimate the heat transport due to AMOC (Watts) by assuming it to be composed of poleward transport of 20 Sv in the upper 1000 m and a return flow below. Assume each of these two flows to carry water of a temperature that represents the average over these two depth ranges in the North Atlantic, as seen in the figure in Box 6.1. Compare to the world energy consumption per second.

   (b) Draw the AMOC estimate based on historical ship observations as a function of year.

   (c) Draw the observed RAPID AMOC time series and a seasonal sine function fit to it.

   (d) Plot the monthly climatology of RAPID and the sine fit. Super-impose the few available AMOC transport estimates based on historical ship observations as if all observations were taken during different months of one year. Discuss your results.

2. **Projections:**

   (a) Plot the RCP8.5 projected AMOC time series of maximum streamfunction value until 2100.

   (b) Contour the AMOC streamfunction averaged over the first decade of the RCP8.5 scenario (2006–2016), as well as averaged over the last decade of this century (2090–2100). Discuss.

3. **Stommel model:**

   (a) Plot the steady states of the AMOC for fresh water forcings of $F_s = 0, \ldots, 5$ m/yr. Discuss.

   (b) Integrate the time-dependent differential equation for ΔS in a scenario of *slowly* increasing fresh water forcing, and plot the resulting AMOC (q) as a function of time.

   (c) Plot $d\Delta S/dt$ as a function of ΔS and discuss the stability of the three possible solutions.

   (d) Explain how tipping points and hysteresis may occur in this model.

4. **Guiding questions to be addressed in your report:**

   (a) What is the AMOC, and why is it important?

   (b) What controls its strength? Why might it change in a future climate? What are the implications of such a change?

   (c) Discuss climate tipping points and explain why the AMOC may be an example.

   (d) What are the consequences of such climate tipping points, in general, in terms of restoring climate to its preindustrial state once CO_2 is reduced back to preindustrial levels?

   (e) What are the consequences of climate tipping points in terms of our ability to predict the occurrence of abrupt climate change and in terms of our ability to gradually adjust to changes?

7

CLOUDS

Minmin Fu and E.T.

Key concepts

- Cloud types: high/low, water/ice, convective/stratiform
- Cloud albedo and greenhouse effects: shortwave (SW) and longwave (LW) cloud radiative effects (CRE)
- How clouds form, moist atmospheric convection
- Cloud microphysics: hydrometeor types, water and ice nucleation, droplet size distribution, droplets fall speed, condensation and evaporation, aerosols
- Cloud feedbacks and warming uncertainty

The discussion in this chapter is very different from that in other chapters. Rather than explaining what changes are observed and projected, and why these changes occur, our focus here is the issue of uncertainty. Clouds and the processes involved in their formation are believed to be a (perhaps even *the*) major reason that the uncertainty range in the projection of the response of the globally averaged surface temperature to a doubling of the CO_2 concentration has essentially not been reduced since the estimated range of 1.5–4.5 °C in the Charney report of 1979. Our objective is to understand why clouds and cloud feedbacks are a source of such uncertainty, and how this uncertainty affects the projected warming magnitude. In the process, we discuss topics such as atmospheric moist convection that are involved in the understanding of other critical global warming issues in other chapters.

Globally, clouds reduce the amount of SW radiation absorbed by Earth by reflecting about 50 W/m^2 on average. On the other hand, clouds trap about 30 W/m^2 of LW radiation and prevent it from escaping to outer space, so that the net effect of clouds globally is a cooling by 20 W/m^2. The processes that lead to cloud formation occur on scales from below a millimeter to kilometers and beyond, and therefore include many small scales that cannot be explicitly simulated in global climate models. Their effects must therefore be indirectly and empirically deduced from larger-scale atmospheric fields, a process referred to as *parameterization*. This leads to considerable model biases and to inter-model disagreement on future changes to the spatial distribution of clouds, and therefore to the projected warming and estimated climate sensitivity (the temperature response to CO_2 doubling; see section 3.1). Our objective here is to understand cloud formation processes, why they are difficult to represent in models, and how they lead to uncertainty in future climate predictions.

In the following sections we review some cloud fundamentals (section 7.1), explain the physics behind the process of moist convection that leads to cloud formation (section 7.2), discuss processes that occur on the droplet scale (*microphysics*, section 7.3), and conclude with a demonstration of how cloud feedbacks affect the response to an increase in greenhouse gas concentration using a simple energy balance model (section 7.4).

7.1 CLOUD FUNDAMENTALS

Clouds are aggregates of water droplets, ice particles, and/or other suspended particles that can be made of both water and ice (mixed-phase), depending on the cloud type and its altitude. The typical sizes of particles in clouds range from 1 to 100 μm in diameter for drops and ice particles all the way up to larger raindrops (500+ μm diameter, or 0.5 mm) or hail (up to a few cm). Typical cloud droplets (say ~20 μm) are considerably smaller than the thickness of a human hair (0.070 mm, or 70 μm), and the typical distance between droplets in a cloud is on the order of 1 mm. Clouds are composed of liquid water when their temperature is above 0 °C, of ice particles below −38 °C, and of either or both phases between these temperatures. Averaged globally, the water vapor present in the atmosphere is sufficient to cover the planet in a water layer roughly 25 mm (about an inch) thick, while cloud ice and liquid water would form a layer of roughly only 80 μm (0.08 mm) thick, 300 times thinner. However, the optical scattering processes associated with droplets mean that clouds, in fact, exert profound radiative effects.

Cloud radiative effect

This is a measure of the effects of clouds on SW and LW radiation. While the details of how cloud radiative effect (CRE) is estimated are complex, we may very crudely describe this procedure as using climate models to calculate the radiation (SW or LW) and then repeating the calculation without clouds. CRE is the difference in radiation (typically calculated at the surface or at the top of the atmosphere) with and without clouds, for example, $\text{CRE}_{LW} = LW_{\text{with clouds}} - LW_{\text{without clouds}}$, where LW is the downward longwave radiation at the surface, for example, measured in W/m^2.

Cloud albedo and emissivity, SW and LW CRE

Because water only weakly absorbs visible (shortwave) radiation (you can see through a glass of water), clouds absorb very little shortwave radiation. They either back-scatter (reflect) or transmit it. Droplets whose size is as small as about 0.5 μm are already very effective at scattering visible light because its wavelength is in the same length scale range as the droplet size. For a given total water content of a cloud, the cloud albedo (and thus its SW CRE) is therefore controlled primarily by the cloud particle size and the number/density of cloud droplets. Clouds such as low stratocumulus have high SW albedo, while thin wispy high clouds such as cirrus clouds have low water content (in the form of ice particles) and thus low SW albedo. Earth's global albedo is roughly doubled by the effects of clouds, from 0.15 due to surface albedo to about 0.3. Clouds reflect very little longwave radiation and either absorb or transmit it. Water and ice droplets in clouds absorb nearly all longwave radiation directed at the cloud because water molecules are effective at absorbing LW (section 2.2). As a result, the atmospheric LW emissivity, which controls the strength of the greenhouse effect, strongly depends on the cloud cover.

High versus low clouds, ice versus water clouds

The greenhouse effect of clouds (as quantified by the longwave CRE) is primarily a function of cloud height rather than water content. First, it takes relatively little water for a cloud to effectively absorb LW radiation, and therefore the LW CRE does not strongly depend on water content. Second, a cloud absorbing LW radiation then re-emits it at the cloud temperature. Low-level clouds (below 2 km), radiating at a temperature close to that of the surface, therefore radiate upward at about the same rate as the ground and thus do not effectively block the LW radiation coming from below. High-level clouds (above 6 km), on the other hand, are very cold, like their surrounding air, given the atmospheric lapse rate that implies a cooling with altitude of 5–9.8 K/km. They effectively absorb the LW radiation from the ground, yet because of their cold temperature radiate upward at a much lower rate and thus have a large greenhouse effect (reflected by their LW CRE). Furthermore, these high-level clouds are composed of ice particles that are larger and fewer than drops in low water clouds and are therefore not as good as water clouds at scattering shortwave radiation. However, they are very effective at absorbing LW radiation. This is the reason high and thin cirrus clouds, composed of ice, are a most effective greenhouse warming climate component.

Low-level water clouds include stratocumulus clouds (lower 1–2 km of the atmosphere) that cover broad regions over the subtropical oceans and have a strong albedo effect. As we will see shortly, deep cumulus clouds, extending over a large vertical extent within the troposphere, involve upward motion of air that then flows out of the cloud horizontally at some height. High-level ice clouds often occur in such outflow regions of deep cumulus clouds. This leads to the formation of flat anvil-like cloud extensions common in the tropical regions, which often lead to the formation of thin wispy cirrus clouds composed of ice particles.

The bottom line is that low clouds are warmer, tend to be composed of water droplets, and have a stronger albedo (net cooling) effect, while high clouds are colder, made of ice particles, and have a strong greenhouse (net warming) effect. The net radiative effect of other clouds, such as mid-level clouds or cumulonimbus clouds that can extend from 1 to 2 km above the surface all the way to the upper troposphere, varies depending on circumstances.

7.2 MOIST CONVECTION AND CLOUD FORMATION

We now wish to understand how clouds form, including via the process of moist convection in the atmosphere, an understanding that will allow us to appreciate the sources of uncertainty in the simulation of these processes in climate models. An appreciation of these processes will also be helpful when we consider the polar amplification of global warming, hurricanes, droughts, forest fires, heat stress, extreme precipitation events, and more.

Atmospheric moist convection and cloud formation

We start with a qualitative description of the moist convection process. Consider a parcel of air lifted adiabatically (that is, with no exchanges of heat or air with the environment) to a higher altitude, where the pressure is lower. The parcel expands and therefore cools. This is the reason that temperature decreases with height within the troposphere (lower 17 km of the atmosphere in the tropics, 6–8 km at mid-latitudes). The maximum possible weight (kg) of water vapor in 1 kg of moist air at a temperature T, the *saturation specific humidity* $q^*(T, p)$, is a function of the temperature and pressure. It is given via the Clausius-Clapeyron relation (Box 2.1) and increases exponentially with the temperature. As the rising parcel cools, $q^*(T, p)$ decreases and the parcel specific humidity (kg moisture in kg moist air) eventually exceeds its saturation value. The parcel then becomes saturated, and water vapor condenses, forming droplets (clouds!). The initial lifting of air parcels leading to this condensation may be forced when a parcel near the surface is carried up by strong turbulent motions in the lower atmosphere, when air flows upward when encountering a cold front, in air flow over mountains, or when air overlying warm surface land or ocean is heated by the surface, becomes buoyant (lighter), and rises. The condensation releases latent heat, heating the parcel and making it more buoyant, forcing it to rise again. This leads to a positive feedback that causes the parcel to continue on its path upward to more condensation and more lifting. The entire process is referred to as *moist atmospheric convection.*

Calculating the temperature and humidity of a rising air parcel

The temperature profile of a lifted parcel, and the corresponding condensation and cloud formation, may be calculated using a simple energy conservation argument. This can be expressed using *moist static energy* (MSE), which is a thermodynamic variable that represents the energy per unit mass of a moist air parcel and is conserved when the parcel is lifted adiabatically in the atmosphere. MSE is given by

$$MSE = c_p T(z) + Lq(z) + gz.$$

The three terms correspond to the *internal energy* $(c_p T)$ due to the parcel temperature $(T,$ measured in K), with c_p (J/K/kg) the specific heat for constant pressure; the potential energy (gz) due to the parcel height $(z,$ m) and gravity $(g,$ m/s^2); and the latent heat of evaporation/condensation (Lq), taking into account the potential to produce heat by condensing the moisture given by the specific humidity q (kg moisture per kg moist air), based on the latent heat of condensation, L (J/kg). MSE conservation is simply a statement of the conservation of energy of an adiabatically raised air parcel.

We now use MSE conservation to calculate the temperature profile of a lifted parcel starting at the surface, at a height of $z = 0$, with a surface temperature T_s, surface specific humidity q_s, and therefore a surface MSE of $MSE_s = c_p T_s + Lq_s$. Typically, the surface air is not saturated, and the surface relative humidity is $RH_s = q_s/q^*(T_s, p_s) < 1$, where p_s is the surface pressure. Initially, as the parcel is lifted to a low level z and cools to a temperature $T(z)$, the saturation specific humidity $q^*(T(z), p(z))$ remains larger than the specific humidity of the parcel,

q_s, and therefore no condensation occurs and the specific humidity does not change. The temperature of the rising parcel can then be calculated by using the fact that its MSE at a level z is equal to its original MSE at the surface, or $MSE(z) = MSE_s$, written explicitly as $c_p T(z) + Lq_s + gz = c_p T_s + Lq_s$. This takes the reduced form $DSE(z) = c_p T(z) + gz = c_p T_s = DSE_s$, where DSE is the *dry static energy* that is conserved in an adiabatic process as long as there is no phase change (evaporation or condensation) of water in the parcel. This conservation of dry static energy leads to the temperature profile for the rising parcel $T(z) = T_s - gz/c_p$, reflecting a cooling with height (known as the *dry adiabatic lapse rate*) of $dT/dz = -g/c_p = -9.8$ K/km.

The rising-induced cooling eventually leads the parcel to a temperature $T(z)$ at which the parcel is saturated, $q_s = q^*(T(z), p(z))$. MSE is still conserved, but upon further cooling some water vapor must condense so that the specific humidity does not exceed its saturation value, $q(z) = q^*(T(z), p(z))$. The temperature is calculated again via the energy conservation statement $MSE(z) = MSE_s$, which now takes the form

$$c_p T(z) + Lq^*(T(z), p(z)) + gz = MSE_s.$$

This is a nonlinear equation for the temperature T, given that the saturation specific humidity depends nonlinearly on T via the Clausius-Clapeyron relation (Box 2.1). As the air parcel continues to rise and cool after the saturation level is reached, the specific humidity $q(z)$ now decreases with height, remaining equal to the saturation specific humidity of the parcel at each level. The reduction in specific humidity reflects a conversion of energy from latent heat to internal energy, which reduces the rate of cooling as the parcel ascends. The rate of cooling with height is now referred to as the *moist adiabatic lapse rate*.

Both the initial unsaturated and later saturated phases of the air-parcel ascent may be calculated by writing the MSE conservation as

$$\text{MSE}_s = c_p T(z) + L \min\left(q_s, q^*(T(z), p(z))\right) + gz \qquad (7.1)$$

and solving this nonlinear equation for the temperature T at each vertical level z. The equation can be solved graphically by plotting the LHS and RHS as a

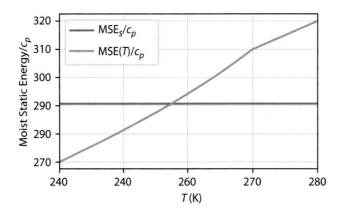

Figure 7.1: A graphical solution of moist static energy conservation.
The RHS and LHS of the moist static energy conservation eqn (7.1) plotted as a function of temperature (divided by c_p so that they have units of temperature Kelvin). The initial parcel is assumed to have a temperature of 280 K and a relative humidity of 80%, and the level to which it is raised, where the energy conservation is evaluated, is $z = 3000$ m. The crossing point is at a temperature $T \approx 258$ K that satisfies the conservation equation and therefore represents the temperature of a moist parcel adiabatically raised to this level.

---

function of T and finding the value of T for which they are equal (Figure 7.1). Alternatively, we can use a root finder routine to calculate T for which the difference between the RHS and LHS vanishes. Note that in equation (7.1), the saturation humidity $q^*(T, p)$ depends not only on the unknown temperature of the parcel but also on the pressure of the parcel's surroundings, $p(z)$, that is calculated next.

The atmospheric vertical pressure profile

The atmospheric vertical pressure profile, $p(z)$, may be approximated by considering the vertical force balance of a parcel of air between the heights of z and $z + \Delta z$. The parcel is pushed upward by the force per unit area due to the pressure at its lower face, equal to $p(z)$, and is pushed downward at its upper part by $p(z + \Delta z)$. The net pressure force per unit area is therefore $p(z) - p(z + \Delta z) \approx -(dp/dz)\Delta z$ and is upward, as the pressure decreases with height. This force may be assumed to a good approximation to be balanced by the weight of the parcel per unit area, given by $-(\Delta z \rho)g$, where ρ is the parcel density, $(\Delta z \rho)$ its mass per unit area, and g the gravitational acceleration. The condition that these two forces sum to zero, so that the parcel does not accelerate upward or downward, is referred to as the hydrostatic balance and gives the hydrostatic equation

$$\frac{dp}{dz} = -\rho g.$$

Using the ideal gas law, $\rho = p/(RT)$, with R the specific gas constant for dry air, $R = 287 \text{ J}/(\text{kg K})$, this may be written as $dp/dz = -pg/RT$. The last equation may be further rewritten as $d\log p = -g/RT dz$. The tropospheric temperature typically varies between 220 K and 300 K, not a large range relative to its absolute magnitude. Setting the temperature therefore to be a constant approximately equal to its averaged value, say $\bar{T} = 260$ K, and integrating the last equation from the surface where $z = 0$ to z, we find $\ln p(z) - \ln p_s = -gz/R\bar{T}$, where p_s is the surface pressure. This yields an exponentially decaying pressure with height,

$$p(z) = p_s e^{-gz/(R\bar{T})},$$

where the decay scale $R\bar{T}/g$ is referred to as the *scale height* of an atmosphere and is roughly 7–8 km on Earth.

Stages in moist convection. The temperature profiles of three parcels starting with different surface RH values are in Figure 7.2, where the assumed atmospheric temperature profile through which the parcels are rising is shown by the dashed line. Consider first the orange curve in Figure 7.2a, corresponding to an initial surface RH of 70% and an initial temperature equal to that of the environment (the dashed and orange lines coincide at the surface). Initially, the adiabatic cooling makes the parcel colder than its environment and hence denser. It must therefore be forced to rise by one of the factors mentioned above. At around 1 km, the parcel reaches saturation and begins to condense (this is the *lift condensation level*, LCL), and its temperature begins to decrease less rapidly with height than before due to the latent heat release. At around $z = 2$ km, where the orange curve crosses the dashed line corresponding to the environmental temperature profile, the parcel is finally warmer and therefore lighter than its surroundings (*level of free convection*, LFC). It can now rise on its own (*free convection*, where the vertical rising does not need to be forced) until the orange and dashed lines intersect once again at around $z = 10$ km (the *level of neutral buoyancy*, LNB), typically condensing all of its water vapor along the way. At the LNB, there is a horizontal outflow from the convecting column. This outflow carries with it some condensed water (in the form of ice particles) and can create extended

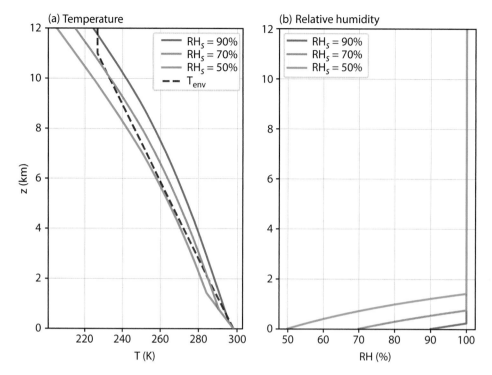

Figure 7.2: Moist convection.
(a) Atmospheric temperature profiles of three air parcels lifted adiabatically from the surface, through an assumed atmospheric environment temperature profile (dashed line). (b) The resulting relative humidity profile of the rising parcels. Parcels are initialized with surface RH of 50%, 70%, and 90%.

surrounding regions of high clouds such as anvil and cirrus clouds that strongly enhance the atmospheric greenhouse effect.

If the initial relative humidity of the parcel is too low (e.g., RH = 50%), there is not enough latent heat of condensation available to make the parcel lighter than its environment (the green line in Figure 7.2 never crosses the dashed line). As a result, free convection never occurs, the vertical rising does not persist, and neither does the condensation and cloud formation.

Uncertainty in representing moist convection. This discussion allows us to understand some of the sources of climate model uncertainty involved in the representation of moist convection and cloud formation. Air parcels in the atmosphere never rise adiabatically, as they tend to entrain air from their surroundings due to turbulence that develops as air rises in moist convection or detrain air into their surroundings as they keep rising. It is very difficult to figure out how much entrainment or detrainment occurs at each level, as the turbulence is characterized by small scales from centimeters to hundreds of meters, far below the resolution of climate models. Additionally, rising and cooling air parcels may be at supersaturation, where the relative humidity is above 100%, and thus latent heat release does not occur as we outlined above. The level of supersaturation depends, for example, on the availability of cloud condensation nuclei and other factors related to cloud microphysics (discussed below) and is thus quite uncertain as well. There are additional such difficulties due to sub-grid scale motions in clouds and atmospheric convection, and the cloud parameterization problem is thus still unresolved in spite of decades of research.

Non-convective clouds

Clouds (that is, stratiform clouds) can also form via several other mechanisms that do not involve atmospheric moist convection, due to either radiative cooling of air parcels that leads to condensation, gradual ascent in warm fronts, and so on. While deep convection occurs mostly in the tropics, over warm land surfaces, and in mid-latitude storms, stratiform clouds can occur, for example, over the mid-latitude and subtropical oceans. Our ability to accurately simulate stratiform clouds and the corresponding feedbacks in global climate models is currently also unsatisfactory.

7.3 CLOUD MICROPHYSICS

Another significant source of uncertainty with regard to the role of clouds in global warming involves the range of processes that occur on the scale of cloud hydrometeors (water droplets, ice crystals, aggregates, graupel, and hail), referred to as *cloud microphysics*. Cloud formation is strongly affected by the presence of aerosols, small liquid or solid particles, either natural (dust, salt particles, volcanic emissions) or human-caused via the combustion of fossil fuels (such as gasoline, oil, and coal), wood, and other biomass. These small atmospheric particles serve as cloud condensation nuclei (CCN) that lead to water droplet formation (nucleation) and as ice nuclei (IN) that lead to ice particle formation. The CCNs and IN allow a much faster condensation and thus a more efficient formation of cloud droplets and ice particles in the presence of aerosols (referred to as *heterogeneous nucleation*) than can occur in their absence (*homogeneous nucleation*). Once formed, these initial droplets grow by deposition of water vapor, followed by collision and coalescence of small particles that can grow all the way to raindrop size and start falling out of the cloud. If the temperature drops below $-10\,^{\circ}$C or so, ice particle formation initiates, again followed by growth by vapor deposition, which can lead to snow formation. When snow particles stick together they form aggregates, and ice particles that incorporate water drops can lead to the formation of graupel and hail. All of these processes must be represented (parameterized) in a climate model even though they occur on a small, unresolved scale, unavoidably leading to model errors and uncertainties.

The uncertainty further increases due to the fact that clouds contain a range of droplet and particle sizes, and it is important that we estimate the number per unit volume of different particle sizes and their phase (water vs ice), as these sizes strongly affect the SW albedo and LW emissivity of clouds. The drop (and other types of particles) size distribution is typically crudely assumed in a climate model to be represented by a certain statistical probability distribution function such as the log-normal distribution. That is, the number of cloud droplets per unit volume between the sizes d and $d + \Delta d$ is assumed to be given by $f(d)\,\Delta d$, where $f(d) = (d\sigma\sqrt{2\pi})^{-1}\exp(-(\ln d - \mu)^2/(2\sigma^2))$. This distribution has two parameters: μ is the average of $\ln d$, and σ is the standard deviation of $\ln d$. Of course, the actual distribution of drop sizes is not guaranteed to follow any known function, and this is another example of a parameterization. The microphysics parameterization needs to determine, for example, how the two parameters μ and σ change due to all microphysical processes, including collision-coalescence, nucleation of new particles, condensation/evaporation, and more. In addition, the microphysics algorithm needs to calculate at any given time the size distribution of each hydrometeor and the conversion between different types. Because these cloud characteristics have a large impact on the radiative properties of clouds and hence on climate, this is an example of processes that occur on a sub-millimeter scale yet have global climate consequences.

Once raindrops form, an additional set of processes must be taken into account. While large raindrops fall rapidly, small droplets remain suspended for a longer time due to frictional drag with the surrounding air. A falling drop will accelerate until friction with air exactly balances the gravitational pull, at which point its velocity does not accelerate, becomes constant, and is referred to as the *terminal velocity*. Small drops (< 0.05 mm) remain spherical as they fall, due to their surface tension, and the air flow around them as they fall is laminar (not turbulent). The terminal velocity is then given by the *Stokes velocity* of a falling spherical object through a viscous fluid, given in this case by $(2/9)gr^2(\rho_{water} - \rho_{air})/\nu$. Here, g is the gravitational acceleration, r is the radius of the assumed spherical drop, ρ_{water} is the density of the drop, ρ_{air} is the density of the surrounding air, and ν (m^2/s) is the dynamic viscosity of air. While a typical raindrop falls at a few meters per second, the terminal velocity of typical cloud droplets (say radius of about 20 μm) is around 1 cm/s. This is a small velocity compared to typical air-rising (updraft) speeds in convective clouds (a few m/s), so cloud droplets remain essentially within the rising air parcels they condensed into. As a result, cloud dissipation can occur via the evaporation of such suspended droplets, in addition to dissipation via precipitation due to the merging of cloud droplets into larger raindrops. This picture is complicated by the fact that the flow around drops larger than 0.1 mm becomes turbulent, and drops larger than 1 mm lose their spherical shape while falling. With non-spherical drops, and with turbulence developing due to the flow around the falling drop, the terminal velocity must be determined numerically or empirically, is therefore more uncertain, and is found in experiments not to exceed about 10 m/s at typical near-surface atmospheric pressures.

The uncertainty in the treatment of microphysics is additionally increased due to the radiative effects of aerosols. Aerosols can both absorb or reflect sunlight (the so-called *direct effects* of aerosols) affecting the atmospheric energy balance. The presence of aerosols affects the level of supersaturation, as well as the number and size distributions of formed cloud droplets and ice particles, and therefore the radiative effects of clouds (the *indirect effect*). Cooling effects due to anthropogenic aerosols may hide some of the greenhouse warming that has occurred so far, making it difficult to estimate climate sensitivity (section 3.1). Because anthropogenic aerosol concentration is enhanced near polluting industrial centers and is thus very inhomogeneous in space, the global effect of aerosols is difficult to quantify. Much of what concerns possible aerosol-climate interactions is still unknown, and aerosols are said to be responsible for a significant part of the climate projection uncertainty.

Much of the above uncertainty relates to challenges climate models face in trying to reproduce present-day climate. The response of clouds to warming poses yet another set of difficulties, as cloud area, geographical and altitude distribution of clouds, and cloud types can change in a warmer climate. Additionally, rain formation processes may change, making the problem of predicting precipitation patterns in a future warm climate highly uncertain, as reflected by the large model disagreement on these processes (section 12.6).

7.4 CLOUD FEEDBACKS AND CLIMATE UNCERTAINTY

As mentioned, globally, the greenhouse effect of clouds currently reduces outgoing longwave radiation by 30 W/m^2, while their albedo effect reduces the absorbed SW by an average of 50 W/m^2, hence leading to a net negative effect (cooling) of around 20 W/m^2. The important question in the context of anthropogenic global warming is how the SW and LW CREs are expected to change, as a relatively small change in clouds will have a large effect, potentially comparable to—or larger than—the direct radiative effect of CO_2 doubling.

How the cloud radiative effect is projected to change is the dominant source of uncertainty in climate models. Cloud feedbacks are often estimated to be positive, meaning that the net global CRE (due to the combined greenhouse LW and albedo SW effects) is expected to become less negative in a warmer climate, leading to further warming. However, the responses of both SW and LW CREs to warming are highly uncertain and involve significant model disagreement.

An example energy balance model with cloud feedbacks

To demonstrate how cloud feedbacks may affect the warming response of the climate system to an increase in greenhouse gases, consider a slightly modified version of the two-level energy balance model introduced in section 2.1.2,

$$C_{\text{surface}} \frac{dT}{dt} = S(1 - \alpha(T)) + \epsilon(CO_2, T_a)\sigma T_a^4 - \sigma T^4$$

$$C_{\text{atm}} \frac{dT_a}{dt} = \epsilon(CO_2, T_a)\sigma T^4 - 2\epsilon(CO_2, T_a)\sigma T_a^4. \tag{7.2}$$

We have added the time rate of change terms on the LHS with $C_{\text{surface,atm}}$ being the heat capacity per m^2 of the lower atmosphere and upper ocean (subscript surface) and of the atmosphere (atm). Cloud feedbacks enter here in two ways: first, via the dependence of the albedo $\alpha(T)$ on surface temperature, representing the dependence of low clouds' properties (e.g., cloud area) on this temperature, and second, via the dependence of the atmospheric LW emissivity $\epsilon(T_a)$ on atmospheric temperature T_a, representing, for example, the dependence of high cloud area, and thus emissivity, on atmospheric temperature. Because these temperatures depend on CO_2, the cloud-related changes to albedo and emissivity act as a feedback on greenhouse warming. These specific assumed dependencies are meant as an illustrative example, as high clouds and low clouds in fact depend on a much larger set of variables, as part of the moist convection process reviewed above. Given these assumptions, the cloud feedbacks may crudely be modeled as

$$\alpha(T) = \alpha_0 \left(1 + \Delta_{SW}(T - T_0)\right)$$

$$\epsilon(T_a) = \epsilon_0(CO_2) \left(1 + \Delta_{LW}(T_a - T_{a,0})\right)$$

$$\epsilon_0(CO_2) = 0.75 + 0.05 \log_2(CO_2/280). \tag{7.3}$$

The parameters Δ_{SW} and Δ_{LW}, which may be positive or negative, introduce a simple linear dependence of the SW albedo and LW emissivity on the surface and atmospheric temperatures, correspondingly. The cloud feedbacks are evaluated as a function of the deviation of the temperature from prescribed reference temperatures T_0 and $T_{a,0}$. The true values of the cloud feedback parameters Δ_{SW} and Δ_{LW} are not known, as different climate models produce different estimates. As an example, suppose that the surface albedo due to low clouds decreases by 4% for a 4 °C of surface warming and the emissivity increases by 0.4% for a 4 °C of atmospheric warming. That means that we need to set $\Delta_{LW} = 0.004/4\,°\text{C}^{-1}$ and $\Delta_{SW} = -0.04/4\,°\text{C}^{-1}$. The base albedo, α_0, and emissivity, ϵ_0, may be set to their values in section 2.1.2. The logarithmic dependence on CO_2 seen in the last line of equation (7.3) was discussed in section 2.2.4.

Figure 7.3a shows the response of the energy balance model to a gradual CO_2 doubling scenario with these assumed cloud feedbacks and without them

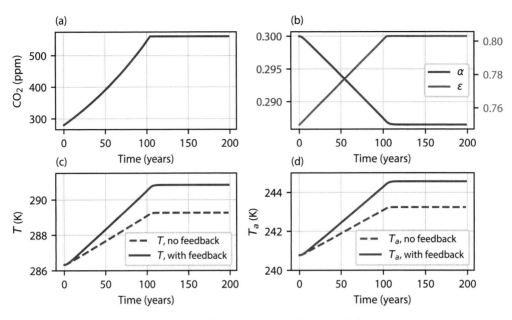

Figure 7.3: Cloud feedbacks in a two-layer energy balance model.
Response of a two-level energy balance model to SW and LW cloud feedbacks. (a) Atmospheric CO_2 as a function of time, representing a doubling scenario. (b) The change to cloud albedo and emissivity resulting from the formulation in eqn (7.3). (c) The surface temperature as a function of time with and without cloud feedbacks. (d) Same, for the atmospheric temperature.

(that is, with $\Delta_{LW} = \Delta_{SW} = 0$). As the temperature is increasing with CO_2, the albedo is decreasing and the emissivity is increasing (Figure 7.3b), so that both represent positive feedbacks with respect to temperature. Note that a seemingly weak dependence of the cloud albedo and emissivity on temperature leads to a significant temperature response, as shown by the difference between the dashed and solid lines in Figures 7.3c,d. This demonstrates the uncertainty involved in cloud feedbacks.

While various negative cloud feedbacks have been suggested in the scientific literature, careful examination by numerous studies has so far not been able to reliably identify a negative cloud feedback that would lead to a significant reduction in the warming expected due to increased greenhouse gas concentration. Climate models seem to consistently underestimate climate sensitivity as deduced from proxy observations of past warm climates (Box 4.1), suggesting that the scenario of negative cloud feedbacks leading to a significantly lower-than-projected temperature increase can be ruled out.

At the same time, much of the difference in the future temperature projections of different climate models can be attributed to their different simulations of the response of low and high clouds to warming, due to different parameterization choices made by different model development teams. While the parameterizations are likely the source for these model differences, and thus for the unsatisfactorily large and unyielding uncertainty in climate projections, the precise reasons for the different cloud responses in different models are difficult to pinpoint and address, precisely because of the many interacting feedbacks involving clouds.

7.5 WORKSHOP

A Jupyter notebook with the workshop and corresponding data file are available; see https://press.princeton.edu/global-warming-science.

1. **Cloud radiative effect and climate sensitivity in climate models:** Plot and compare the results of two prominent climate models (from the Geophysical Fluid Dynamics Laboratory, Princeton, NJ, and the Hadley Center, UK), and discuss the implications to global warming uncertainty. First, clouds (three contour plots per each of the two models):

   (a) The global distribution of cloud fraction at year 1850.

   (b) The projected global distribution of cloud fraction at year 2100, according to the RCP8.5 scenario.

   (c) The difference between the two.

   Next, temperature and CRE (three contour plots for each of the two models):

   (d) The global distribution of the change in surface air temperature (SAT) between the preindustrial state (often referred to as historical and representing year 1850) and the RCP8.5 scenario at 2100: $\Delta SAT = SAT_{RCP8.5} - SAT_{historical}$.

   (e) Change in LW CRE: $\Delta CRE_{LW} = CRE_{LW,RCP8.5} - CRE_{LW,historical}$.

   (f) Change in SW CRE.

2. **Convection and cloud formation:** Consider an atmospheric vertical temperature profile with a prescribed lapse rate of 6.5 K/km and an air parcel starting at the surface with a temperature equal to that of the atmospheric profile there and at 70% saturation.

   (a) The parcel is lifted from the surface to $z = 3$ km. Plot the MSE conservation (LHS vs RHS of equation 7.1) as a function of T, and find the parcel's final temperature as the one for which the LHS minus the RHS vanishes. Is the parcel saturated at that point?

   (b) Plot the parcel's profiles of temperature and relative humidity as it rises for $z = 0, 0.1, \ldots, 10$ km. Describe the results: At what heights are the LCL? LFC? LNB?

   (c) Qualitatively (no equations): If the above rising air parcel gradually entrains dry air from its surroundings so that it doubles its volume by the time it is at 1 km, what do you expect the consequence to be for its temperature, RH, LFC, and its rising motion?

(d) ***Optional extra credit:*** Calculate the profiles of temperature and relative humidity of an air parcel as it is lifted while entraining environmental air. Assume the parcel entrains a fraction 0.1 of its volume per kilometer and that the environmental air has a lapse rate of 6.5 deg/km and an RH of 50%.

3. **Cloud feedbacks in a two-level energy balance model:** Global mean surface temperature is predicted to increase by 2–4 °C due to doubling of CO_2.

(a) Calculate the steady state atmospheric and surface temperatures for $CO_2 = 280$ without cloud feedbacks; use the solution as the initial conditions as well as the reference temperatures in the cloud feedback terms. Now run a warming scenario with CO_2 increasing exponentially on a timescale of 150 yr until doubling, and continuing at a double CO_2 concentration. Repeat this run first without and then with cloud feedbacks, using $\Delta_{LW} = 0.001$ and $\Delta_{SW} = -0.01$. Plot the time series of CO_2, T, T_a, emissivity, and albedo for the runs with and without cloud feedbacks. Discuss the effects of cloud feedbacks on climate sensitivity (equilibrium surface warming due to $\times 2CO_2$) in this case.

(b) Estimate the SW cloud feedback parameter Δ_{SW} needed to *increase* the predicted warming due to the CO_2 doubling by 2 °C. To do this, tune Δ_{SW} until the difference in temperature between 560 ppm and 280 ppm is now 2 °C higher than the difference in temperature with Δ_{SW} set to zero. Next, estimate the percent change in albedo between 560 ppm and 280 ppm with the calculated Δ_{SW}. Discuss the sign of this calculated parameter.

(c) Estimate the LW cloud feedback parameter Δ_{LW} needed to *decrease* the predicted warming due to the CO_2 doubling by 2 °C by tuning Δ_{LW}. Next, estimate the percent change in emissivity between 560 ppm and 280 ppm with the calculated Δ_{LW}. Discuss the sign of this calculated parameter.

4. **Guiding questions to be addressed in your report:**

(a) Why are clouds important in the current climate? Why are they important in the context of global warming?

(b) How do clouds form and dissipate, and what are the uncertainty sources in representing them in climate models?

(c) What are the mechanisms by which clouds can play a role in affecting the response to increased concentration of CO_2?

(d) What are the uncertainty sources in identifying and quantifying each of the mechanisms by which clouds affect the response to increased CO_2?

(e) Should we not worry about global warming because of the large uncertainty due to clouds?

HURRICANES

Key concepts

- The big factors: sea surface temperature (SST), wind shear
- Have hurricanes become stronger already?
- A possible future intensification: Clausius-Clapeyron relation, hurricane energetics, potential intensity

Hurricanes that occur over the Atlantic Ocean and northeast Pacific Ocean, and typhoons over the northwest Pacific Ocean, involve rapidly rotating winds around a low pressure center. Hurricanes are characterized by sustained winds over 119 km/hr, with major hurricanes (categories 3–5) having sustained winds over 178 km/hr. More generally these storms are referred to as tropical cyclones, and they occur in most tropical oceans. About 15–25 hurricanes affect the United States in a given decade; some recent ones (e.g., Katrina

2005, Sandy 2012, Harvey 2017) have caused most significant damage, and there are reasons to suspect these storms may get stronger in a warmer climate, although whether the signal is already detectable is still being debated.

In the following sections, we try to understand how hurricanes form and what are the factors that determine their strength (section 8.1), how one understands the dependence of the maximum possible hurricane strength on the warming sea surface temperature (potential intensity, section 8.2), what can be said about any observable changes, and what are the factors that currently limit our ability to confidently project changes to hurricanes in a future warmer climate (section 8.3).

8.1 FACTORS AFFECTING HURRICANE MAGNITUDE

Hurricane development

African easterly waves are the source of most Atlantic hurricanes. These are localized weather systems that travel off the west coast of Africa into the tropical Atlantic Ocean, some of which develop into tropical storms and even hurricanes. Once over the warm tropical ocean, these disturbances develop a low pressure center, with winds rotating anticlockwise around the low pressure and rising air motions in the center (Figure 8.1). Because evaporation from the ocean surface is enhanced by stronger surface winds, the wind involved in these disturbances leads to enhanced surface evaporation from the tropical ocean. The resulting warm and moist air then rises in the center of these weather systems, and due to the decrease of pressure with altitude the rising air expands and therefore cools. The cooling leads to the saturation of the rising air parcels, according to the Clausius-Clapeyron relation (Box 2.1), and therefore to condensation, releasing latent heat of condensation and heating the atmosphere, and the heating further strengthens the rising motions (section 7.2).

In a developed hurricane, the center of the storm forms an *eye* with a diameter of 30–60 km that is clear of clouds and calm, surrounded by the *eye wall* with rapidly rotating cloud clusters and containing the strongest winds in the hurricane. The surface air flowing toward the storm center rises within the eye wall region, and the rising air is replaced by a horizontal air flow near the surface toward the center of the developing cyclone. The Coriolis force due to Earth's rotation (section 4.2.1) forces the horizontal inflow required to replace the rising

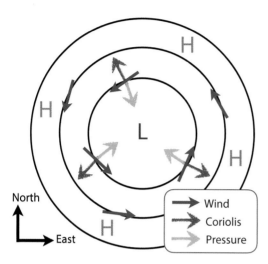

North
East

	Wind
	Coriolis
	Pressure

Figure 8.1: Schematic of a low pressure system in the Northern Hemisphere.
Winds (blue arrows) are flowing anticlockwise, mostly along constant pressure lines (black circles). The pressure force (green arrows pointing from high to low pressure) and the Coriolis force (red arrows to the right of the wind) balance so that the wind is in a geostrophic force balance (section 4.2.1). Near the surface, the friction with the ocean also plays a role, making the winds deviate slightly toward the low pressure center. These converging motions toward the center low pressure drive rising motions at the center of the system. In the case of intense tropical cyclones the centrifugal force also plays a role in this force balance, supplementing the outward pointing Coriolis force and creating what is known as a *gradient balance* between the pressure force on the one hand and the centrifugal and Coriolis forces on the other.

air to the right (in the Northern Hemisphere), strengthening the anticlockwise rotational motion around the developing storm (Figure 8.1). This leads to a positive feedback: the stronger the surface wind, the stronger the evaporation and therefore the stronger the heating due to condensation in the rising air. The stronger heating leads to more intense flow of rising air, therefore faster surface inflow toward the storm center, which further strengthens the rotational surface winds via the Coriolis force. This closes the positive feedback cycle. The positive feedback strengthens the initial weather perturbation into a tropical storm, possibly forming a hurricane. In the process, heating due to the condensation of water vapor is converted into horizontal wind energy. The strength of this positive feedback is controlled by two main factors, the sea surface temperature (SST) and the wind shear.

Sea surface temperature

A warm SST leads to a warmer surface air via air-sea heat exchanges. The maximum possible amount of water vapor in the surface air (e.g., the saturation specific humidity, kg water vapor per kg of moist air, which is a function of temperature and pressure, $q^*(T, p_s)$, where p_s is the surface atmospheric pressure) increases exponentially with the temperature according to the Clausius-Clapeyron relation (Box 2.1). This moisture later condenses in rising motions in the center of the hurricane and leads to atmospheric heating, which fuels the hurricane. The SST can therefore control the maximum possible hurricane strength via its effect on the near-surface atmospheric moisture content. As saturation moisture increases exponentially with temperature, the SST can be a significant factor controlling hurricane development. The latent heat flux lost by the ocean during the evaporation that later leads to the atmospheric heating is the ultimate source for the atmospheric heating. Thus, a high upper ocean heat content, a result of a deep and warm surface mixed layer (Box 6.1), is also an important factor that allows strong hurricanes to develop.

Wind shear and the role of El Niño

Vertical wind shear is either a change of horizontal wind *speed* with height, or a change in wind *direction* with height, or both. If strong enough, it can prevent a hurricane from developing or weaken a well-developed hurricane. While the mechanism by which vertical wind shear disrupts a hurricane is complex and not completely understood, we briefly mention two specific processes that have been suggested. First, the moist convection and latent heat release that drive the hurricane winds can be weakened by a *ventilation* by dry air that is brought in from the environment by the shear. This dry air can flow either into the hurricane core at mid-level or into the boundary layer (lower 1–2 km) air that flows toward the hurricane center. Second, the magnitude of the central low pressure can be weakened via a dilution of the upper level warm core of the storm by mixing it with the cooler surrounding air, again driven by the shear.

The wind shear over the tropical Atlantic is affected, among other factors, by El Niño events in the eastern equatorial Pacific (Box 8.1). The warm equatorial east Pacific SST during such events leads to stronger upper atmosphere westerly winds and stronger lower level easterly winds over the tropical Atlantic, together increasing the vertical wind shear. As a result, El Niño events correspond to a weaker hurricane season over the Atlantic. Similarly, La Niña events, with their cooler east equatorial Pacific SST, lead to a weaker wind shear over the tropical Atlantic and to a stronger hurricane season there. Figure 8.2a shows

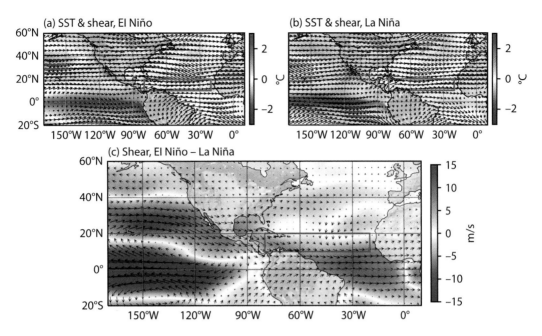

Figure 8.2: El Niño and wind shear effects on hurricane development.
The North Atlantic MDR is indicated by the green boxes. (a) SST anomaly from monthly climatology (that is, the deviation from monthly averages) during El Niño events, shown by colors, in °C. Also shown are anomaly wind shear vectors $(\delta U, \delta V) = (U(200\,\text{hPa}) - U(850\,\text{hPa}), V(200\,\text{hPa}) - V(850\,\text{hPa}))$ during El Niño events, where U is the zonal wind and V the meridional wind. (b) Same, for La Niña events. (c) The shear $(\delta U, \delta V)$ during El Niño events minus that during La Niña events is shown by the vectors. Colors show the magnitude of the shear difference in m/s, $\left((\Delta U)^2 + (\Delta V)^2\right)^{1/2}$, where $\Delta U = \delta U_{\text{El Niño}} - \delta U_{\text{La Niña}}$.

the SST anomaly (colors) and the wind shear (vectors, defined as the wind at 200 hPa minus that at 850 hPa) averaged over El Niño events. Panel b shows the corresponding SST and wind shear vectors for La Niña events. Finally, panel c shows the difference in shear magnitude between El Niño and La Niña events as color shading and the corresponding shear difference vectors as arrows. Note in particular the *main development area* (MDR) for Atlantic hurricanes shown by the green box at 10–20°N, 80–20°W; the box in Figure 8.2c contains mostly red, indicating that the shear is stronger during El Niño events.

8.2 POTENTIAL INTENSITY

The mechanism of hurricanes has been described as a *heat engine* that converts heat to the kinetic energy of the strong hurricane winds. The source of heat is latent heat released by the condensation of water evaporated from the ocean, and the sink of kinetic wind energy is mostly the dissipation due to friction of hurricane winds with the ocean's surface. Using this framework one can estimate the maximum expected wind speed of hurricanes, and how it might change in a warmer climate.

Energy/power dissipation in a hurricane

The dissipation of energy in a hurricane is largely due to friction between surface winds and the ocean. In general, the work done by a force is equal to the force times the distance over which the force is applied. The energy dissipated by the wind friction force per unit time is therefore the wind friction force times the distance, divided by the time it takes the winds to cover that distance, or the friction force times the wind velocity. The friction force per unit area between the winds and the surface is estimated as $C_D \rho_{air} V_s^2$, where V_s is the surface wind speed (m/s), C_D (nondimensional) is an empirical *bulk constant*, and ρ_{air} is the surface air density (kg/m^3). Multiplying by the wind velocity, we obtain the energy dissipation per unit area per unit time, $D = C_D \rho_{air} V_s^3$, and the integral of this quantity over the hurricane area is the total energy dissipation per unit time. A time series of this quantity is referred to as the power dissipation index (PDI). The work done by the hurricane wind friction force on the surface is also a measure of the destructiveness of the hurricane, making the PDI a useful measure of the expected damage caused by hurricanes.

Climate Background Box 8.1
El Niño, La Niña

These are the warm (panel a of the accompanying figure) and cold (b) phases of the El Niño–Southern Oscillation (ENSO) climate variability mode in the equatorial Pacific, which occurs due to a large-scale interaction between the ocean and atmosphere. *Southern Oscillation* refers to a variability of the sea level atmospheric pressure difference between Tahiti (in the Pacific) and Darwin, Australia, which co-varies with the sea surface temperature in the eastern equatorial Pacific. The time series

(c) shows the SST anomaly (deviation from monthly mean) averaged over what's known as the NINO3.4 index region, 120°W–170°W, 5°S–5°N, shown by the green box in panels a and b. Panel a (b) shows the El Niño (La Niña) SST composite—that is, the SST anomaly averaged over all months during which the index is above 1 std (below minus 1 std). The superimposed vectors show the corresponding wind anomaly composites. The time series shows warm El Niño events in red and cold La Niña events in blue, with warm events occurring irregularly every 2–7 years.

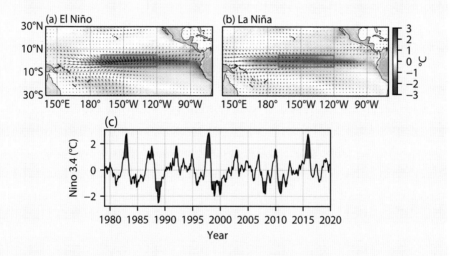

The two phases of ENSO have significant and different effects on weather, affecting precipitation, temperature, and winds over large areas of the globe. In spite of their irregular occurrence, these events can now be predicted a few months in advance, providing advance warning to farmers, fisherman, and other interested parties around the equatorial Pacific and elsewhere, that are affected by this variability. The simulation of ENSO by climate models is getting better, although predictions of how it might change in a future climate still seem very uncertain due to sensitivity to model formulation and differences between the projections of different models.

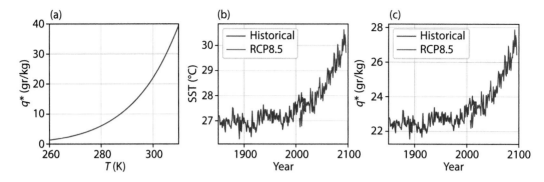

Figure 8.3: SST and saturation specific humidity over the Atlantic main development area.
(a) Saturation specific humidity as a function of temperature, at a pressure of 1000 hPa. (b) Historical and projected SST in the main development area of Atlantic hurricanes according to the RCP8.5 scenario. (c) Saturation specific humidity based on the SST in panel b.

Saturation specific humidity response to a warming SST

As a reminder, the saturation specific humidity, $q^*(T, p)$, is exponential in the temperature (Box 2.1; Figure 8.3a). When the temperature is 30 °C the saturation specific humidity is 27.5 gr/kg, and a mere one degree of warming leads to a very significant 6% increase in the saturation specific humidity. Figures 8.3b,c show time series of the observed (historical) SST averaged over the main development area of hurricanes in the Atlantic, and the predicted SST there according to the RCP8.5 scenario, as well as the expected resulting large increase in saturation specific humidity. We next estimate how the anticipated increase in saturation specific humidity for a warmer SST may affect the projected maximum possible hurricane strength.

Energy input into a hurricane

Water evaporated from the ocean surface into a hurricane is transported upward in the hurricane eye wall surrounding its center. The rising air expands and cools at the lower pressures it encounters and this leads to condensation. The condensation involves the release of latent heat, fueling the hurricane. Ultimately, this heat is drawn from the warm upper ocean, which is cooled by the evaporation process and which is the main source of heat for the hurricane. The evaporation per unit area, per unit time, into the hurricane is calculated as $C_k V_s \cdot (q^*(T, p_s) - q_a)$, where V_s is the surface wind speed; T is the surface air temperature (K) assumed approximately equal to the SST; $q^*(T, p_s)$ is the saturation specific humidity at the surface (in units of kg moisture per kg of moist air); q_a is the atmospheric surface specific humidity; and C_k is a nondimensional empirical *bulk coefficient* for calculating evaporation. This evaporation leads to a heat input into the atmosphere via latent heat release upon condensation of $L \rho_{air} C_k V_s \cdot (q^*(T, p_s) - q_a)$, where L is the latent heat of evaporation/condensation (J/kg), and ρ_{air} the surface air density (kg/m^3). The fraction ϵ of this heating converted into kinetic energy of the winds is the *efficiency* in transforming heat to kinetic energy (further discussed below). The net input of kinetic energy into the hurricane due to evaporation, per unit time, is, therefore,

$$G = \epsilon L \rho_{air} C_k V_s \cdot (q^*(T, p_s) - q_a) = \epsilon L \rho_{air} C_k V_s \cdot q^*(T, p_s)(1 - RH),$$

where $RH = q_a / q^*(T, p_s)$ is the atmospheric relative humidity at the surface, and we note that the energy input G is a function of the SST via the dependence of the surface saturation moisture on temperature.

Efficiency of converting heat to wind energy

A heat engine converts heat into mechanical (kinetic plus potential) energy, and a four-stage heat engine was imagined by Carnot as a piston filled with gas and attached to a reservoir of heat. By heating and cooling the gas in the piston, it is made to expand and contract, turning the heat input into work done by the piston, which produces mechanical energy. As part of this Carnot cycle of heating and cooling, heat is provided to the heat engine at a high temperature T_H, some of the heat is converted to mechanical energy, and some is lost to the environment at a low temperature T_C due to the imperfect efficiency of the engine. The maximum possible efficiency, defined as the mechanical energy extracted divided by the heat provided to the engine, can be shown to be $\epsilon = (T_H - T_C)/T_H$. For hurricanes in this analogy, the produced energy is the kinetic energy of the hurricane winds, and $T_H = SST$ is the temperature of the heat source (the ocean surface). T_C is the average temperature at which heat is lost by the hurricane via LW radiation to outer space by air at the top of the storm. The taller a hurricane is, the lower the temperature T_C at its top and, thus, the greater the thermodynamic efficiency. For a typical hurricane, with $SST \approx 30\,°C \approx 303$ K and $T_C \approx 200$ K, we therefore find the maximum expected efficiency to be $\epsilon \approx 1/3$.

Hurricane as a Carnot cycle

To justify the use of Carnot efficiency to estimate the energy conversion efficiency in a hurricane, consider the flow of air in a hurricane, along the surface toward the center, up in the eye wall near the storm center, outward at the top of the storm, and back down toward the surface, as an analog of the compression and expansion of gas in a piston in a Carnot heat engine. Stage 1 of the Carnot cycle involves isothermal expansion of the gas-filled piston while absorbing heat from the external reservoir, countering the adiabatic cooling that would have occurred due to the expansion. This corresponds in a hurricane to air acquiring heat (in the form of latent heat from surface evaporation) as it flows along the surface around and toward the center of the storm, at a warm surface temperature T_H. Stage 2 is the adiabatic expansion of the piston, with no heat exchanges with the external reservoir, doing work and cooling. This is analogue in a hurricane to air rising up near the storm center, releasing latent heat and cooling adiabatically. Stage 3 is isothermal compression, in which the piston motion compresses the air while releasing heat to an external cold reservoir to counter the adiabatic heating due to compression. In a hurricane, this is analogue to the release of heat by the air in a hurricane via radiation while flowing out of the convective plumes at the top of the storm, descending, and therefore compressing at a temperature T_C. Stage 4, the final step, is adiabatic compression while heating the air in the piston, which is parallel in a hurricane to air descending back to the surface and adiabatically heating.

Estimating maximum wind speed in a hurricane

Setting dissipation to be equal to the energy input ($D = G$), we find $V_s^2 = \epsilon L q^*(T, p_s)(1 - RH)C_k/C_D$. Assuming the ratio of the two bulk coefficients to be about one,

$$V_s^2 = \frac{SST - T_C}{SST} L\, q^*(T, p_s)\,(1 - RH). \tag{8.1}$$

This is the final expression for the potential intensity (V_s) we have been looking for, representing the squared hurricane wind speed on the left in terms of

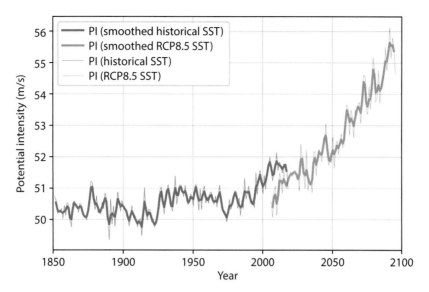

Figure 8.4: Potential intensity.
PI calculated using eqn (8.1) from the main Atlantic hurricane development area SST averaged over the hurricane season, observed and predicted according to RCP8.5.

the important factors: (1) evaporation, which depends on the saturation specific humidity and on the surface atmospheric relative humidity; (2) conversion of evaporated water to heat as expressed by the constant L; and (3) efficiency of conversion of heat to kinetic energy as expressed by the ratio involving the warm and cold temperatures within the hurricane. The hurricane velocity estimated this way is exponential in the sea surface temperature because of the Clausius-Clapeyron relationship that affects surface evaporation, the energy source for hurricanes in this picture.

Substituting values for the different constants, assuming the tropical surface atmosphere to be 85% saturated, $RH = q_a/q^*(T, p_s) = 0.85$, and setting the surface temperature to $T = 25\,^{\circ}\text{C}$, we find a reasonable estimate for present-day hurricane wind speed,

$$V_s = \left(\frac{1}{3} \times (2260 \times 10^3) \times \frac{20}{1000} \times (1 - 0.85) \right)^{1/2} = 47\,\text{m/s} = 169\,\text{km/hour}.$$

Figure 8.4 shows the estimated wind speed calculated based on the observed and predicted RCP8.5 SST in the Atlantic main development area. Some small strengthening of this potential intensity wind speed is already expected due to the warming experienced by the MDR (blue line), and more is anticipated by year 2100 (orange line). Of course, the anticipated increase in potential intensity is subject to possible modifications by any changes that might occur to the shear due to El Niño and to a number of other factors; we discuss below the issue of ACC detection in actual observations of hurricane intensity.

Caveat

It is important to remember that, while elegant and powerfully simple as a concept, potential intensity provides only an *upper bound* on the strength of hurricanes. That bound depends on quantities, such as the efficiency of converting heat to kinetic wind energy discussed above, that are difficult to estimate because a hurricane is clearly not an ideal heat engine, and this efficiency may change in a warmer climate. Even taking potential intensity as an accurate measure of the maximum possible hurricane intensity, the actual strength of individual hurricanes depends on factors beyond the sea surface temperature and the heat content of the upper ocean. Hurricanes are weather systems, and their development depends on complex interactions within the atmosphere (the shear discussed above being but one example), preventing many storms from reaching their maximum capacity.

8.3 OBSERVED CHANGES TO HURRICANE ACTIVITY

One naturally wonders if the hurricane intensification expected based on the above arguments is already observed. As there is a significant year-to-year variability in the number and intensity of hurricanes, it is not obvious that one can detect the signal of anthropogenic climate change from the short observed record at this point. Figure 1.3c shows that it is difficult to detect a change in the number of hurricanes per year. We next consider two further approaches to the detection of climate change signal in the hurricane record.

We first attempt to correlate the observed PDI with the observed SST in the main development area of hurricanes during the hurricane season in the North Atlantic. Figure 8.5 shows the two time series, as well as a smoothed version of each. There are some similarities between the PDI and SST, yet the correlation r^2 is only 0.18 for the non-smoothed data. A similar result is found if using only

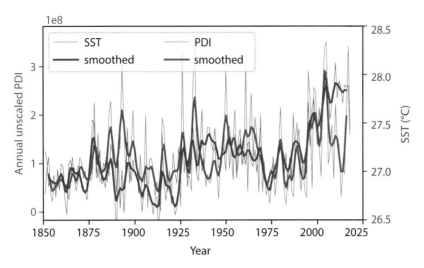

Figure 8.5: Power dissipation index and SST.
Observed annual time series of the PDI (red) and SST (blue) in the Atlantic main development area during the hurricane season.

the data starting with 1950, which are more recent and therefore more reliable. For the smoothed time series, the correlation is 0.31 and is again similar when using only the more recent data starting in 1950. It should be noted that while one might expect the intensity of hurricanes to increase with SST following the above arguments based on potential intensity, PDI is not a direct measure of hurricane intensity, as it also reflects the number and lifetime of storms, neither of which is predicted by potential intensity. It is therefore difficult to predict with certainty what the dependence of PDI on SST should be.

As an alternative second approach, we calculate a linear (regression) line fit to a specific measure of the intensity of hurricanes as a function of time to see if we can identify a statistically significant increase. For this purpose, consider data of wind intensity, given every 6 hr, for each observed hurricane, and calculate the fraction of hurricane intensities that are of category 3 and above (and are thus considered major hurricanes). The line fit in Figure 8.6 shows an increase in the global fraction of major hurricanes at a few percent per year. The increase is statistically significant ($p < 0.01$), yet with $r^2 = 0.18$, indicating that (only) 18% of the variance of the signal is explained by the linear trend. A similar picture is seen for the North Atlantic, but it is not possible to identify a significant increasing trend in other ocean basins.

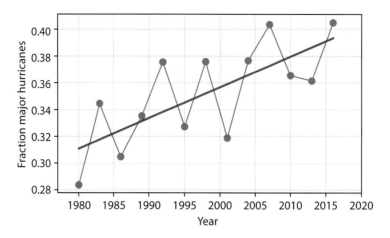

Figure 8.6: Observed changes to major hurricanes.
The fraction of hurricane winds that are in category 3 and above, as a function of year, with data points binned to three-year averages.

Given that the warming over the ocean so far has been only about a degree Celsius, the mechanistic explanation we discussed above, involving potential intensity, does not predict a large increase in hurricane intensity at this point. It is therefore perhaps not surprising that the signal is difficult to extract from the noise due to natural variability in hurricane number and strength. Thus while economic damages by hurricanes have increased over time, it is important to remember that socioeconomic factors, such as increased population density along coastlines, play a significant role in this increase. Further uncertainties in our ability to predict future hurricane strength are introduced by factors such as wind shear that need to be predicted, complicated in turn by the difficulty to predict changes to El Niño events in a warmer climate.

The inherent uncertainty in finding trends in hurricane characteristics using the short and noisy record has been dealt with in recent studies by separately considering type I and type II errors for detection and attribution. Making a type I error means that anthropogenic climate change was *wrongly* found to contribute to a change in hurricanes. A type II error means that the analysis concluded that ACC is not responsible for the change when in fact it did contribute significantly. The science requires, of course, that type I errors be avoided. Policy implications may in some cases require that type II errors be avoided by taking into account predicted risks not justified on a purely scientific basis.

8.4 WORKSHOP

A Jupyter notebook with the workshop and corresponding data file are available; see https://press.princeton.edu/global-warming-science.

1. **Observations of the number of Atlantic hurricanes and of the PDI:**

   (a) Number of hurricanes per year: Plot the time series of the estimated number of Atlantic hurricanes. Add a linear regression line to examine whether there is a trend, calculating the p and r^2 values, and discuss your results.

   (b) Plot the time series of the PDI, and of the SST averaged over the Atlantic MDR, on the same axes (using two separate y axes).

   (c) Smooth the two time series using two passes of a 1-2-1 filter,

   $$\overline{T}_i = 0.25T_{i-1} + 0.5T_i + 0.25T_{i+1}.$$

   Note that every pass requires eliminating the first and last data points, where the smoothing cannot be evaluated. Plot the smoothed time series as thick lines over the non-smoothed ones, shown as thin lines. Write the expression for the twice-smoothed temperature $\overline{\overline{T}}_i$ in terms of T_i. Over how many years is the average $\overline{\overline{T}}_i$ calculated?

   (d) Correlation of SST and PDI: Calculate the correlation between the time series of the SST and PDI with and without smoothing for 1950 to 2005. Repeat using the data up to present-day. To calculate the correlation of the two time series, first remove their mean and then calculate the correlation by programming the explicit formula for correlation. Check your results using the NumPy function for calculating correlation. How would you justify calculating the correlation for the twice-smoothed data?

2. **Potential intensity:**

   (a) Plot the saturation specific humidity as a function of time based on the observed MDR SST and the SST increase projected by 2100 under the RCP8.5 scenario.

   (b) Calculate and plot the expected potential intensity as a function of time for the observed (historical) MDR SST and for the projected MDR SST increase by year 2100 under the RCP8.5 scenario (plotting the raw PI, with the smoothed time series superimposed).

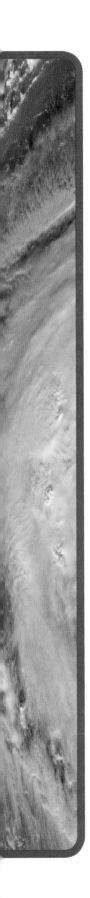

(c) Calculate and plot time series of the cube of the PI to approximate the PDI, and normalize by the mean of the PDI calculated from the historical SST. What is the expected percent increase in PDI by the end of the century? Remember that PDI is a measure of hurricane destructiveness.

3. *Optional extra credit:* **Detecting ACC in hurricane intensity:** Calculate the fraction of major hurricanes every year and plot it as a function of time. Repeat after averaging in bins of three years. Calculate and plot a linear fit with and without the binning, and calculate the r^2 in each case. Discuss your results.

4. **Guiding questions to be addressed in your report:**

   (a) Have hurricanes (as quantified by the PDI, number of hurricanes per year, and fraction of major hurricane winds) already been getting stronger?

   (b) What are the factors affecting hurricane strength?

   (c) How can the maximum possible (potential) hurricane strength be estimated from the SST? Why is it only a *potential* maximum?

   (d) Will hurricanes get stronger in an RCP8.5-like global warming scenario? If so, by how much? (Consider both hurricane wind velocity and destructiveness.)

   (e) What are the sources of uncertainty?

ARCTIC SEA ICE

Key concepts

- Recent changes to Arctic sea ice extent, area, volume, age
- Why these changes occurred, what is the impact
- Sea ice feedbacks: albedo, age and melt ponds, thickness and insulation, thickness and mobility due to storms
- Detection of climate change
- Future projections

The annual minimum of Arctic sea ice area, typically occurring in September, has declined dramatically over the past decades, from 5.8 million square km in 1979 to 2.4 million square km in the record low year of 2012, a decrease of over 50%; see the time series in Figure 9.1a. Sea ice, especially in the marginal ice zone

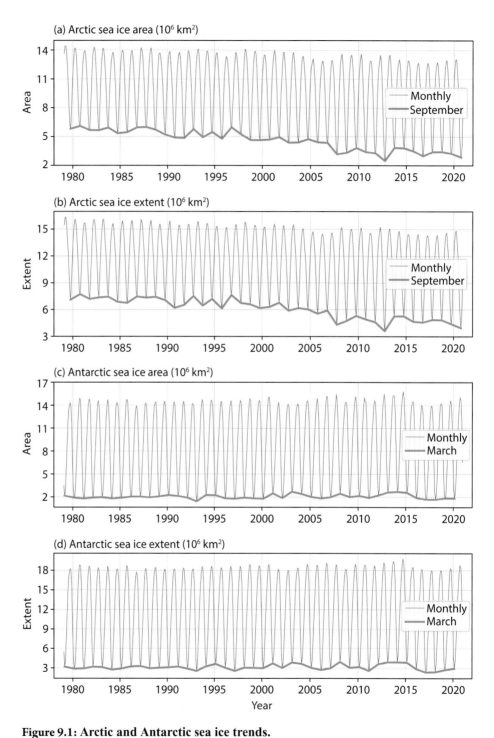

Figure 9.1: Arctic and Antarctic sea ice trends.
Time series of Arctic sea ice area (a) and extent (b), and for Antarctic sea ice area (c) and extent (d).

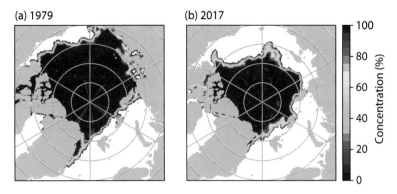

Figure 9.2: September sea ice concentration in 1979 and 2017.

at the edges of the Arctic sea ice area, is composed of individual *floes* whose sizes range from a few centimeters to hundreds of kilometers. The analysis of sea ice cover is often done by dividing the ocean area into grid boxes (say 50 by 50 km). The percent area covered by sea ice in a given grid box is the sea ice *concentration*. The September concentration maps for 1979 and 2017 in Figure 9.2 again demonstrate a dramatic decline. Sea ice *extent* is the area (say in millions of km²) of the region where sea ice concentration is at least 15%. Sea ice extent therefore contains areas of both sea ice and open ocean and is by definition larger than or equal to the sea ice area. The extent measure also shows strong decline over the past decades (Figure 9.1b). This sea ice decline occurred over the satellite era since 1978 and is therefore well observed. In a striking contrast to the Arctic, Antarctic sea ice has not shown any significant decline so far (Figure 9.1c,d). While many mechanisms have been suggested to account for this discrepancy between the recent Arctic and Antarctic trends, the reason for this difference is still not well understood.

The changes to Arctic sea ice cover have significant effects on the local ecology, from plankton blooms to disruptions of resting, breeding, and hunting for Arctic ice-obligate mammals such as polar bears and seals. Indigenous people may be affected via the possible change to Arctic storminess due to the effect of sea ice loss on air-sea fluxes. Sea ice protects shores from large waves, and its reduction therefore leads to coastal erosion and coastal damage. The thinning of sea ice also affects the safety of travel and sea ice–based hunting; on the other hand, it opens new Arctic shipping routes.

The significant reduction of Arctic sea ice area may also affect the mid-latitudes, including temperature, precipitation, and weather events. While these effects are mostly still difficult to distinguish from natural atmospheric variability, they are beginning to be detectable and are expected to be more so with the anticipated further reduction of summer sea ice over the next few decades. Rapid sea ice melting and freezing have been implicated in past abrupt climate change such as the Dansgaard-Oeschger (D-O) events, dramatic warmings in the northern North Atlantic that occurred roughly every 1500 yr during the last ice age (20–60 kyr before present). Each event led to a regional warming of about 10 °C, lasted hundreds to a thousand years, was triggered rapidly within a decade or two, and terminated within a similarly short time. Sea ice melting near Greenland is a likely explanation of these past events, possibly triggered by the variability of the Atlantic meridional ocean circulation (chapter 6). This reinforces the possibility of significant future climate effects induced by sea ice melting, although one needs to keep in mind that D-O events occurred at a glacial time when sea ice cover was much more extensive than today, and thus its melting had an especially dramatic effect on the atmospheric temperature around the North Atlantic.

Apart from its area, concentration, and extent, Arctic sea ice is characterized by thickness and age, which have both changed dramatically over the satellite period (Figure 9.3). Arctic sea ice thickness used to vary from up to a meter for

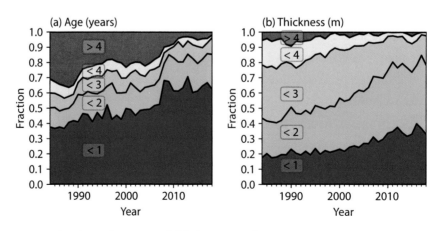

Figure 9.3: Arctic sea ice age and thickness trends.
Time series of the estimated fraction of sea ice area that is covered by different (a) age and (b) thickness categories.

newly formed first-year ice to near 10 m for multi-year ice. With the reduction of ice area, the age and thickness of sea ice have also reduced dramatically. Much of the Arctic sea ice is now below 3 m thick, and it is mostly no older than two years. This indicates that present-day Arctic sea ice is more mobile due to its lower thickness, and it either melts or is exported from the Arctic before growing in age and thickness. We will see below that sea ice albedo varies with ice area, age, thickness, and snow cover, so that the changes to sea ice characteristics over the past decades significantly affect its role in climate.

9.1 PROCESSES AND FEEDBACKS

How sea ice grows and melts

Sea ice starts growing as small (3–4 mm) crystals suspended in the upper few centimeters of the ocean, known as *frazil ice*. In a calm ocean, these later coagulate into *grease ice* and then freeze into a smooth-bottomed sea ice sheet. In a rough ocean, the frazil ice crystals accumulate into *pancake ice*, composed of roughly circular floes from 30 cm to 3 m, shaped by collisions between individual floes. Once covered by a first snow or exposed to cooling, the pancakes merge into a thin layer of rough-bottomed ice that then grows in thickness, leading to older and thicker ice via freezing from below and snow accumulation from above. Melting can occur via melt pond formation from above or via heating and melting by the underlying ocean.

Sea ice area and the albedo feedback

Sea ice albedo (fraction of shortwave radiation reflected) ranges from 0.5–0.8. Its large-scale mean value is affected by sea ice concentration and extent and by the existence of snow cover, which increases the reflectivity relative to that of exposed ice, among other factors. A smaller sea ice area leads to more SW absorption by the ocean near sea ice floes, which leads to further melting, creating a powerful *positive feedback* that can accelerate melting.

Ice age and roughness, and the albedo feedback

The uneven surface height of sea ice, especially of older sea ice, implies that during the melt season, melt ponds accumulate in depressions in the sea ice surface.

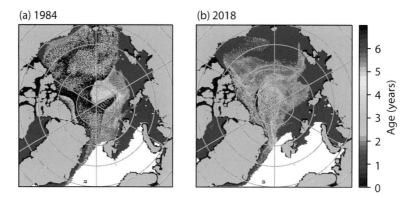

(a) 1984   (b) 2018

Age (years)

6
5
4
3
2
1
0

Figure 9.4: Arctic sea ice age.
Estimated sea ice age in 1984 and 2018, showing a significant loss of perennial (multi-year) sea ice over much of the Arctic.

Melt ponds have lower albedo than the ice itself and therefore absorb more SW radiation, warm, and further melt the ice underneath. This can further deepen the depressions, and as a result, older sea ice, which has been subjected to this process over several years, is characterized by a rougher surface. Melt ponds that form over such older and rougher-surface ice during the summer therefore tend to be deeper and consequently of smaller area for a given volume of meltwater. The effective ice albedo is thus higher for older and rougher ice. As a result of this process, when ice age declines as it has over the satellite era (Figure 9.4), its surface becomes less rough. This results in shallower and larger-area melt ponds and therefore in a smaller effective albedo. This leads to further melting and to an even smaller ice area, amplifying ice retreat via this *positive feedback*.

Ice thickness and insulation feedback

The winter atmosphere above sea ice is typically much colder than the ocean water underneath the ice, which is at its freezing temperature of $T_f = -1.8\,^\circ\mathrm{C}$. As a result, heat diffuses through the ice from the relatively warm ocean toward the cold upper sea ice surface. Because the ocean just under the sea ice is already at the freezing temperature, this heat flux away from the ocean at the base of the sea ice can lead to freezing. However, ice is a good heat insulator, and therefore thicker ice leads to a weaker heat flux away from the ocean below the ice and toward the cold atmosphere. This weaker heat flux away from the ocean under thick ice implies a slower freezing under thicker ice. The upward diffusive heat flux F from the relatively warm ocean to the colder atmosphere is proportional to the vertical temperature gradient in the ice, with the proportionality constant being heat conductivity in ice (κ, in $\text{W m}^{-1}\text{K}^{-1}$) as given by

$$F = -\kappa \frac{\partial T_{\text{ice}}}{\partial z} = -\kappa \frac{T_{\text{surface}} - T_f}{h}.$$

The negative sign after the first equality represents the fact that temperature in the ice increases downward while the heat flux is upward. In the second equality we estimate the temperature gradient within the sea ice layer with the temperature at the surface of the ice minus the freezing temperature at its bottom, divided by the ice thickness h. Thus, a thicker sea ice cover implies less upward heat flux from the ocean, or in other words slower growth of thickness. Clearly this is a *negative feedback* on ice thickness growth as the thickness growth is slower for thick ice. The rate of latent heat release per unit area due to freezing at the ice base, in terms of the ice thickness growth rate, is $\rho_{ice}L_f dh/dt$, where the ice density is ρ_{ice} and the latent heat of freezing is L_f. If we equate that to the heat flux diffusing upward through the ice away from the ice bottom, we find an equation for the sea ice thickness of the form

$$\rho_{ice}L_f \frac{dh}{dt} = \kappa \frac{T_f - T_{\text{surface}}}{h}.$$

The negative feedback on sea ice thickness growth is apparent from the inverse dependence of the thickness rate of growth (LHS) on the ice thickness h (RHS).

Feedback due to ice thickness, mobility, and Arctic storms

Sea ice has been thinning for a few decades (Figure 9.5), possibly since the 1960s, and once thin enough, it is prone to being broken by wind and ocean waves and being transported out of the Arctic by atmospheric storms. This leads to a mostly thin new ice the following year, again prone to being exported, presenting a *positive feedback* on ice thinning. The strong sea ice minima of 2007 and 2012 were forced, among other factors, by unusual atmospheric weather patterns that led to strong Arctic sea ice export.

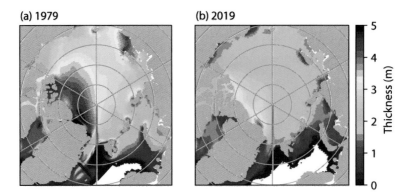

Figure 9.5: Arctic sea ice thickness.
Estimated sea ice thickness in 1979 and 2019, showing overall thinning and the nearly complete loss of ice over 3–4 m thick.

Heat budget of melting sea ice, implications to attribution

This chapter's workshop suggests a calculation, in watts per meter squared, of the amount of heat needed to warm ice to the melting point and then the amount needed to melt it. The relatively small persistent heat flux required over a decade to melt a typical sea ice cover implies a difficulty of attribution of sea ice melting to specific external forcing. The presence of strong positive feedbacks suggests that an initial thinning and melting of sea ice (whether due to an increase in greenhouse radiative forcing or due to a natural variability signal) can be further amplified by the positive feedbacks reviewed above, leading to a dramatic sea ice melting. Regardless of the initial trigger of the current Arctic sea ice decline, given the polar amplification in observed and projected warming (section 3.2), it seems unlikely that Arctic sea ice will fully recover in the near future. The fact that Antarctic sea ice did not show a significant decline over the past decades (Figure 9.1c,d) further adds to the difficulty in attribution of the observed Arctic sea ice changes.

Arctic halocline, warmer subsurface water, and ocean heat transport

The Arctic Ocean layer just under the sea ice is fresh and therefore light in spite of its cold temperature. Underneath, at a depth of only a few tens of meters, one finds the denser, relatively warm and salty water entering from the Atlantic Ocean (related to the NADW; see Box 6.1). The fresh water on top prevents the warmer water from reaching the ice, but a possible change in ocean stratification due to a future climate change can lead to a direct contact of the warm and salty water with the sea ice and to a very rapid melting. With decreasing Arctic sea ice area, a stronger Arctic Ocean cyclonic gyre may develop due to exposure of the ice-free ocean to wind forcing. This may again bring the warm subsurface waters toward the surface and lead to a strong melting, presenting another positive feedback that can lead to Arctic sea ice melting. The transport of subsurface warmer ocean water from low latitudes seems to have played a role in the Arctic melting so far and may be amplified as the ocean warms, playing an even more dominant role.

9.2 DETECTION OF CLIMATE CHANGE

A simple statistical method of detection of ACC can be used to test if the recent sea ice melting trend over the past four decades is likely to be due to natural variability. For this purpose, we calculate the probability distribution function (PDF) of sea ice trends over different periods, using a long *control run* of a climate model, with CO_2 concentration set to its preindustrial value with no increase. A segment of the model time series showing a naturally occurring downward trend of sea ice area is shown in Figure 9.6.

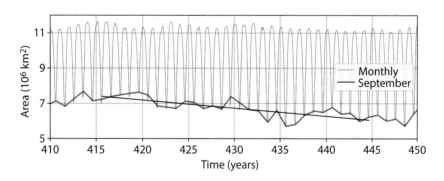

Figure 9.6: Simulated sea ice trend due to natural variability.
Results of a long control run (with CO_2 fixed at its preindustrial value) of a climate model, showing sea ice area during a natural variability event of sea ice loss over an example period of 30 years.

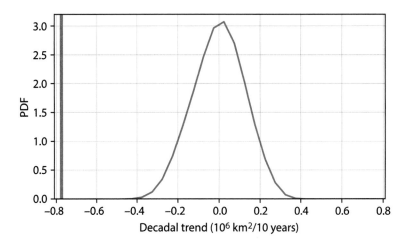

Figure 9.7: Detection and attribution of sea ice trends.
The probability density function of multi-year trends in the annual minimum sea ice area, as a function of the trend magnitude, when examining intervals of 41 yr in a long run of a climate model. The observed trend, over a similar interval, is marked by the vertical gray bar.

The PDF is calculated for a given time interval of, say, N years by scanning the long model sea ice area time series and calculating the trends over all N-year segments. A trend is calculated by fitting a regression line in the form of $area = a \times year + b$ to the September sea ice area time series in each N-year interval. The slope, a, is used to calculate the decadal trend, $T = 10a$. These decadal trends are then binned into preselected trend amplitude bins, and the number of occurrences in each bin is recorded to calculate the PDF as a function of decadal trend amplitude T and interval length N, $f(T, N)$. The PDF is normalized such that $\int_{-\infty}^{\infty} f(T, N)dT = 1$, and the probability of encountering a trend between the values of T and $T + \Delta T$ over a period of N years is then given by $f(T, N)\Delta T$. The calculation is repeated for different time intervals N, from two to a hundred years, to obtain the PDF as a function of time interval and trend amplitude.

The PDF for 41 yr intervals is in Figure 9.7, with the observed trend over the first 41 yr of the satellite record indicated by the vertical gray bar. The probability that the observed trend is due to natural variability is the integral of the area under the curve to the left of the bar, which is less than 0.01. Thus, the probability that the observed trend is due to natural variability rather than climate change is estimated as being less than 1%. This means that we have detected a climate change in the Arctic sea ice area trend at a confidence level of over 99%. The

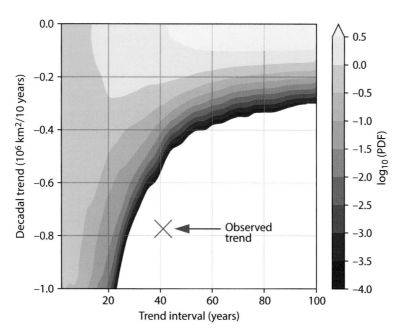

Figure 9.8: Detection and attribution of sea ice trends.
The probability density function of multi-year trends in the annual sea ice minimum area, as a function of the trend magnitude (showing only trends that correspond to sea ice decline) and the interval length, in a long run of a climate model. The observed trend is marked by a red × symbol.

PDF as a function of both trend interval in years and the amplitude of the trend is shown by the contour plot of Figure 9.8, with the observed trend noted by the red ×.

This analysis assumes, of course, that the sea ice variability in the model simulation is realistic, an assumption that is impossible to verify given the short satellite-era observed record. If the model significantly underestimates the magnitude of the variability, then our confidence level may decrease significantly.

9.3 FUTURE PROJECTIONS

Summer Arctic sea ice melting

Remarkably, climate models failed during the first decade of the 21st century to predict the severity and speed of the recent observed September Arctic sea ice melting, even when observations showed it to already be quite prominent.

Current projections under the RCP8.5 scenario are that summer sea ice might mostly disappear before mid-century.

Winter Arctic sea ice melting

While September sea ice is projected to disappear around the middle of this century, winter sea ice is expected to continue to regrow every fall. However, at some point toward the end of the 21st century or by mid-22nd century, the winter maximum sea ice in March is projected by climate models to vanish under the RCP8.5 scenario, and the Arctic may become sea ice–free year-round. The albedo feedback is not negligible during the time of maximum winter sea ice in March as the insolation then is significant, yet it plays a smaller role in the dark Arctic winter-time. The winter sea ice melting is therefore likely assisted by other positive feedbacks. One possible candidate is the initiation of winter-time atmospheric convection and the associated formation of winter Arctic clouds (section 7.2), with a strong cloud greenhouse effect that would keep the Arctic Ocean from rapidly cooling during winter and prevent sea ice from regrowing after the summer. Another possible mechanism is enhanced winter-time downward LW radiation due to a warmer and therefore moister lower atmosphere as moisture increases the atmospheric LW emissivity and the atmospheric warming, of course, increases the downward LW radiation.

9.4 WORKSHOP

A Jupyter notebook with the workshop and corresponding data file are available; see https://press.princeton.edu/global-warming-science.

1. **Observations over the past decades:**

    (a) Time series: Plot the monthly sea ice area and extent for the Arctic and for Antarctica, and superimpose a time series of the values during the month of minimum sea ice (September for the Arctic).

    (b) Concentration: Compare the sea ice September minima of 1979 versus 2017 by plotting the two sea ice concentration maps.

    (c) Thickness: Plot March sea ice thickness maps for 1979 and 2019. Calculate and plot the PDF (probability distribution function) of ice thickness for these two years for the entire year.

    (d) Age: Plot maps of sea ice age 1984 and 2018 for March. Calculate and plot the PDF of ice age for these two years.

2. **How much heat is needed to melt sea ice?** Calculate how much heat per unit area (W/m^2) is needed during the three summer months every year to completely melt a sea ice cover within 30 yr. Assume that the initial ice thickness is 3 m, that it has an initial average temperature of $-3\,°C$, and that the ocean mixed layer (the upper 50 m) is initially at the freezing temperature ($-1.8\,°C$). At the end of the period, the ocean mixed layer is at a final temperature of $T_{final} = 1.2\,°C$ and the ice is completely melted.

    (a) Calculate separately the heat flux required for (i) warming the ice to its melting temperature of $0\,°C$, (ii) melting the ice, (iii) warming the melt water to T_{final}, and (iv) warming the mixed layer to T_{final}.

    (b) Find out the Arctic summer downward SW, and calculate what percent change in cloud or ice albedo during the peak summer months could account for a change in absorbed solar flux that can lead to the above complete ice melting in 30 yr. Discuss the implications for identifying the cause of Arctic sea ice melting.

3. **Increasing roughness with age, melt ponds, and albedo:** (i) Explain why ice surface roughness is expected to increase with ice age. (ii) Suppose the sea ice surface height as a function of location x, y is given by $h(x, y) = h_0 \cos(kx)$ with $k = 2\pi/100\ \text{m}^{-1}$. Plot the area fraction covered by melt water as a function of the melt water volume

for $h_0 = 0.5$ m and for $h_0 = 1$ m. (iii) Based on your results, explain the positive feedback leading to more melting due to the decrease in ice age and therefore decrease in ice roughness.

4. **Projections:** Contour the September sea ice concentration at the beginning of the 21st century and at year 2100 according to the RCP8.5 scenario.

5. **Climate change detection:** (i) Examine if the trend in annual minimum sea ice area over the satellite period since 1979 is consistent with natural variability, as estimated using a long run of a climate model with no anthropogenic greenhouse forcing. Calculate the PDF for a period corresponding to the length of the observed record, and calculate the probability that the observed trend is part of natural variability. Repeat the calculation using only the record from 1979 to 1999; what differences do you see? (ii) Calculate the trends as a function of both the trend duration (2–100 yr) and amplitude. Contour the PDF, and superimpose a symbol showing the observed trend. Discuss.

6. **Animation!** Create an animation of September sea ice concentration maps for all years of the RCP8.5 projection.

7. **Guiding questions to be addressed in your report:**

   (a) Describe the changes in Arctic sea ice over the past decades.

   (b) Why is the Arctic sea ice cover important?

   (c) Why do these changes happen? Discuss all the positive and negative feedback mechanisms you can list and their possible roles in the observed changes.

   (d) Can a climate change be detected? Can the observed trend be attributed to anthropogenic global warming? Explain the difference between detection and attribution in this case.

   (e) What are sources of uncertainty in ACC detection? In the attribution to greenhouse forcing?

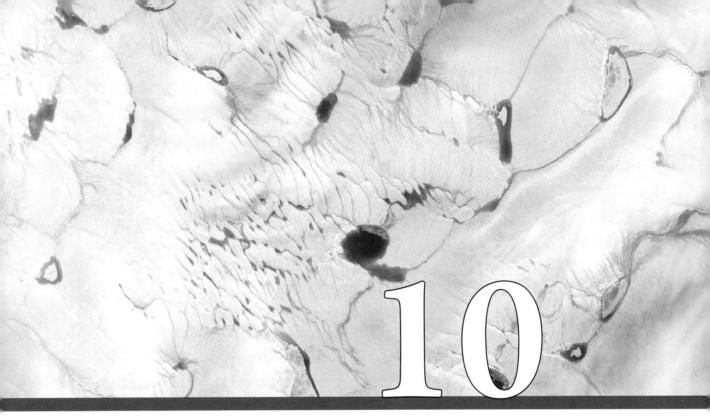

10

GREENLAND AND ANTARCTICA

Wanying Kang and E.T.

Key concepts

- Observed changes to Greenland and Antarctica
- Surface mass balance (SMB): ablation versus accumulation, positive degree days (PDD), elevation-desert effect, lapse rate and reduced ablation, temperature-precipitation feedback
- Calving: yield stress, floating criteria, hydrofracturing in ice shelves
- Marine ice sheet instability (MISI)
- Basal heat budget and melt water production

- Ice streams acceleration, lubrication by basal water, melt ponds, and moulins
- Ice ages
- Future projections

The Antarctica and Greenland ice sheets contain the equivalent of about 70 m and 7 m of global sea level rise, respectively. These ice sheets have existed for millions of years and did not exist during the hot-house climate of the Eocene (56–33.9 Myr), when CO_2 is believed to have been significantly higher than at present (Box 4.1). While there is evidence of ice over Greenland for possibly up to 18 Myr, other evidence suggests that this ice sheet may have melted and regrown over the past few millions of years, when CO_2 was not much higher than at present, suggesting that the Greenland ice sheet may be quite sensitive to warming. There are concerns that Greenland melting is already accelerating and that the West Antarctic ice sheet (also about 7 m sea level equivalent, and whose base is mostly below sea level, making it more vulnerable) may become unstable and melt under a warmer future climate. West Antarctica may have been losing ice during the past few decades, although this may be a result of natural variability rather than anthropogenic climate change. Ice streams that carry ice from Greenland to the ocean have accelerated significantly in recent decades, and summer surface melting seems to occur earlier and to expand to record ice surface areas and magnitudes. These concerns are amplified by the existence of several positive feedback mechanisms that can dramatically accelerate ice sheet collapse.

The timescales of glaciological processes that can lead to significant changes in the massive Greenland and Antarctica ice sheets are very long, on the order of hundreds to thousands of years; the uncertainty of projections necessarily increases with the timescale. Current estimates are that the contribution of ice sheets to sea level rise by the end of the 21st century will likely be up to a few tens of centimeters. But the potential for surprises is there, and there is a clear need to anticipate and understand longer-term risks. Our purpose in this chapter is to understand the processes that can lead to significant changes to the Greenland and Antarctica ice sheets and, based on this understanding, develop an appreciation of the level and specific sources of uncertainty. We start by defining some basic glaciological terms (section 10.1); we follow with a review of processes that contribute to the ice mass balance of these ice sheets, from surface snow accumulation and melting to ice flow (section 10.2); and we finish with a review of observations of the state of these ice sheets over the past decades and with some projections for the 21st century (section 10.3).

10.1 TERMINOLOGY

Consider some basic terms related to the ice mass balance and dynamics of ice sheets that are used later in this chapter to discuss their response to global warming.

Accumulation: Snow fall over an ice sheet that is compacted into ice by the weight of further snow layers.

Calving: Mechanical detaching (as opposed to melting) of ice from the main body of a glacier, often at a water edge where it can form icebergs.

Ablation: Any process leading to ice loss, including melting, evaporation, sublimation, calving, avalanches, ice flow, and wind-driven snow blowing.

Surface Mass Balance (SMB): Rate of accumulation minus surface ablation processes (m/yr).

Ice stream: Rapid (> 1 km/yr, relative to about 1 m/yr ice velocity away from ice streams), narrow (order 50–100 km) ice flow toward the ice sheet edge, embedded in larger ice sheets. The term is also used for outlet (valley) glaciers that drain inland ice sheets or ice caps, whose width is a few to 50 km and which flow at velocities larger than 100 m/yr.

Ice shelf: Floating thick (200 m–2 km) ice that is connected to an ice sheet and is often fed by a flow from an ice stream.

Grounding line: The line separating the grounded part of an ice sheet from the floating ice shelf, where the ice sheet loses contact with the ground and becomes a floating ice shelf.

Marine-based ice sheet: One whose land base is under sea level. Alternatively, a marine ice sheet sometimes refers to one that terminates in the ocean and has a grounding line.

Equilibrium line altitude (ELA): The altitude of the *equilibrium line* separating the zones of net ablation in the lower parts of an ice sheet or glacier and net accumulation at higher altitudes.

10.2 PROCESSES

In equilibrium, the total accumulation over an ice sheet equals the total ablation, and ice flow carries ice from the accumulation zone to the ablation zone, maintaining a steady state ice elevation in both. Consider processes responsible for accumulation, ablation, and ice flow; how they affect the ice sheet equilibrium; and how they may change in a warmer climate.

10.2.1 Accumulation

Snowfall tends to increase with the amount of water vapor in the atmosphere, which is bounded by the saturation moisture, following the Clausius-Clapeyron relation (Box 2.1). The exponential dependence of the saturation moisture on air temperature indicates a possibly strong decrease in snow accumulation due to the atmospheric drying with decreasing temperature. As a result of this dependence on temperature, given that air temperature falls with altitude (by 6.5–9.8 K/km), one expects that the higher an ice sheet is, the less surface accumulation it will receive. This is referred to as the *elevation-desert effect*, which can limit the height of ice sheets. As an example, note that Antarctica is effectively a desert, receiving a very small amount of snow accumulation, both due to its high elevation and thus cold temperature and due to the distance of its vast interior land mass from the source of moisture due to ocean evaporation. The same dependence of moisture on temperature means that climate warming can lead to increased snow accumulation, referred to as the *temperature-precipitation feedback*. This assumes, of course, that the warming does not bring the ice surface to above the melting temperature and does not lead to a change from snow to rain, as both effects will lead to increased ablation rather than increased accumulation.

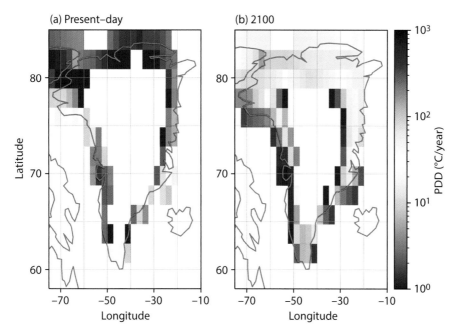

Figure 10.1: PDD over Greenland.
Estimated for present day (left) and for year 2100 based on the RCP8.5 scenario.

10.2.2 Surface melting and PDD

Positive degree days

Melting and sublimation occur mostly during summer, when the ice surface temperature is higher. It is found empirically that melting is then proportional to a measure referred to as *positive degree days*, calculated as the sum of daily-averaged atmospheric surface temperatures (in K) above the melting temperature,

$$\text{PDD} = \sum_i (T_i - T_m) \mathscr{H}(T_i - T_m). \qquad (10.1)$$

Here T_i is the daily-averaged temperature at day i, $T_m = 273.15$ K ($0\,^{\circ}$C) is the melting temperature, and $\mathscr{H}$ is the Heaviside function, that is equal to one when its argument is greater than zero, and to zero otherwise. As an example, four days with temperatures in degrees Celsius of 3, 0, 1.5, and -1 would correspond to a

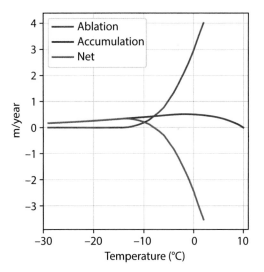

Figure 10.2: Surface mass balance versus temperature.
Schematic of accumulation (blue), surface ablation (red), and net surface mass balance (green) as a function of surface atmospheric temperature.

total PDD of $3 + 1.5 = 4.5$. Figure 10.1 shows an estimate of PDD over Greenland for the present-day versus for an RCP8.5 projection for year 2100, showing a significantly enhanced expected melting at the margins of Greenland.

A schematic of accumulation, surface ablation, and the net surface mass balance as a function of temperature is shown in Figure 10.2. For cold temperatures, snow accumulation (blue curve) increases with temperature due to the temperature-precipitation feedback. When temperatures increase above freezing, precipitation turns into rain, which may flow off the ice sheet as runoff, reducing accumulation. Surface ablation (red) is essentially zero for cold temperatures, rapidly increasing for higher temperatures. The net accumulation (accumulation minus ablation, green line) therefore initially increases with temperature and then rapidly decreases. It is generally believed that Antarctica is in the regime where an increase in net accumulation is expected with an initial future warming, while Greenland is in a regime where its net accumulation will decrease in a future warming.

Albedo feedback and melt ponds

Summer melting of an ice sheet can be accelerated by the formation of melt ponds (also referred to as *supraglacial lakes*) shown in the image on the title page of this chapter. These ponds can develop in response to higher PDD values in a warmer climate and have a low albedo ($\sim$0.1), significantly less than that of snow ($\sim$0.8) or ice ($\sim$0.5). The low albedo leads to SW absorption, further melting, and further growth of the melt ponds. On a much longer timescale, if the ice sheet retreats and contracts, its overall albedo effect also declines, leading to an atmospheric warming and therefore to further melting and ice sheet retreat.

10.2.3 Calving

Crevasses (deep cracks in a glacier, ice sheet, or ice shelf) can initiate when a moving ice sheet stretches due to flow over curved topography or due to downstream flow acceleration that pulls the ice apart. Crevasses near the end of an ice stream lead to calving. Calving at the edge of an ice sheet or ice shelf can also occur when the weight of the edge of an ice cliff that terminates the ice sheet/shelf is larger than the ice strength, often weakened by the presence of crevasses. A dominant mechanism of calving in Antarctica today is the separation of large tabular icebergs, which can be up to tens to hundreds of kilometers large, due to the horizontal propagation of rifts.

Ice strength and yield stress

Ice strength, measured by the *yield stress*, τ_c, is about 100 kPa (where Pa = $\mathrm{kg\,m^{-1}s^{-2}}$). It is the maximum force per unit area that can be sustained by ice before breaking. Consider a vertical ice cliff with a protruding portion (Figure 10.3a). The weight of the protruding section exerts a downward force along the connecting dashed line, equal to $\rho_{ice}LWHg$, where W is the width of the section (into the plane of the figure), L and H the dimensions seen in the figure, ρ_{ice} the ice density, and g gravity. The maximum force that the ice can sustain is the yield stress times the area connecting the protruding section, $\tau_c WH$.

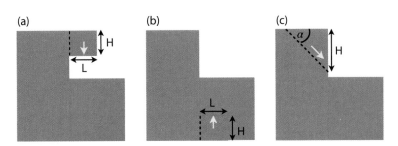

Figure 10.3: Calving.
Three highly idealized scenarios of calving at the edge of an ice shelf.

Thus, the section will break if

$$\rho_{ice}LWHg > \tau_c WH$$

which may be written as a condition on the maximum length of the protruding section,

$$L > \frac{\tau_c}{\rho_{ice}g} = \frac{100 \text{ kPa}}{900 \text{ kg/m}^3 \times 9.8 \text{ m/s}^2} = 11 \text{ m}.$$

A related scenario can occur when buoyancy forces exceed the ice strength and break a submerged part of the ice sheet, as shown schematically in Figure 10.3b and discussed in this chapter's workshop. Finally, *cliff instability* is the process described in Figure 10.3c, where the force parallel to the diagonal dashed line, due to the weight of the ice above that line, is larger than the ice strength, leading to a collapse (*slumping*) of the ice edge (*cliff*), thus limiting the height above water level of such vertical cliffs (see workshop). The thinning of ice via slumping makes the remaining ice prone to breaking by buoyancy forces due to the adjacent ocean that can now lift and break it, causing calving of the entire ice column (full-depth calving). Whether cliff instability is currently an important ablation process is still an open issue. More generally, calving mostly occurs on a small scale that cannot be resolved by models of entire ice sheets. This necessitates a parameterization, much as atmospheric moist convection, clouds, and microphysics need to be parameterized in atmospheric models, or as small-scale ocean mixing needs to be parameterized in an ocean model. The need for a parameterization implies a large uncertainty involved in the quantification of the role of specific calving mechanisms in the fate of large-scale ice sheets in a warming climate.

Hydrofracturing

Once crevasses are initiated at the surface of an ice shelf or near an ice sheet edge, they can fill with surface melt water. This can cause the crevasses to penetrate deeper, as the water's density is higher than that of ice, and the water acts as a heavy wedge pushing into the ice. Hydrofracturing happens, and the water-filled crack can then propagate downward, when the column mass of water is sufficiently deep that its weight overcomes the resistance of the ice material strength to developing fractures and the opposing local ice pressure. The resulting cracks enhance calving so that a continuous surface melt water production can accelerate calving and ice loss beyond its direct effect on surface mass loss. This is one process by which climate warming can accelerate mass loss by ice sheets. Hydrofracturing was implicated in the dramatic collapse of the Larson B ice shelf over a short period of a few weeks in 2002, captured by a remarkable series of NASA satellite images.

10.2.4 Ice flow

The Antarctic ice sheet maintains a steady state mass via snow accumulation in its interior balanced by ice flow toward the ocean. The slow ice flow in the interior of the ice sheet is driven by the surface height gradients between the thick interior and thinning ice sheet edges toward the ocean. This slow flow then feeds narrower and faster ice streams that flow toward the ocean, balancing much of the interior snow accumulation. Ice stream flow is opposed by friction at both the ice base (basal friction) and with the surrounding ice (lateral friction). The ice flow is also resisted by a back-pressure force by the grounded front edge of the ice stream and by the ice shelves sometimes attached to this front edge.

Calving rate at the front edge of an ice stream both depends on the rate of upstream flow toward the ice terminus and strongly affects this rate, as follows. On the one hand, calving rate is controlled by the rate at which ice flows toward the calving zone, and in particular across the grounding line. At the same time, calving at the ice edge, and ice shelf breakup (e.g., via hydrofracturing or rift propagation), can lead to reduction in the back-pressure force and to ice flow acceleration.

The total ice mass transport across the grounding line, Q, can be shown to be proportional to the fifth power of the thickness, H, of the ice at the grounding line, $Q \propto H^5$. The fifth power depends on the exponent $n = 3$ from *Glen's law*, which describes the nonlinear dependence of the ice deformation rate (given, e.g., by the flow shear $\partial u / \partial z$, where u is the horizontal ice velocity and z the vertical coordinate) on the force (stress) acting on the ice. The fifth power also depends on various other assumptions that involve significant uncertainty. In any case, the strong dependence on ice thickness implies that a thicker ice at the grounding line would end up transporting mass away from the ice sheet at

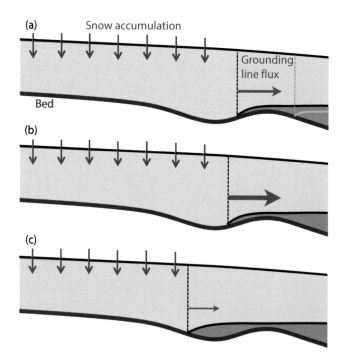

Figure 10.4: A schematic of the **marine ice sheet instability**.

a much faster rate. This is the basis for the marine ice sheet instability (MISI) described next, which is one mechanism that may lead to a rapid collapse of an ice sheet and to rapid sea level rise.

Marine Ice Sheet Instability (MISI)

Consider an ice sheet (Figure 10.4a) whose grounding line was located at the marked green line and which retreated upstream due to melting induced by, say, ocean warming to the location shown by the vertical dashed line. Suppose that this new location is in an equilibrium in the sense that snow accumulation on top is equal to the ice mass flux across the grounding line. As a result of the retreat, the grounding line is now at a location where the bedrock (brown curve) slopes upward toward the ocean (a *retrograde* slope). It turns out that this new location is unstable and leads to further retreat as follows. Suppose the new upstream location of the grounding line further retreats by a small amount, as in Figure 10.4b. The bedrock elevation is now lower than before; thus the ice thickness is larger at the grounding line. As a result of the fifth-power dependence of the grounding line mass transport on the ice thickness, the ice transport across the grounding line (red arrow) would be even larger. The larger outflow of ice across the grounding line leads to faster calving, and the ablation rate is now larger than the accumulation rate, causing a net ice loss and therefore a further retreat of the grounding line. This positive feedback, referred to as MISI, would lead to a continued retreat of the grounding line until it arrives at a location where the bed slopes downward toward the ocean (*prograde* slope; see Figure 10.4c). The new location is stable to this process: further retreat of the grounding line would lead to a thinner ice at the grounding line because of the higher topography there. This decreases the mass flux across the grounding line, which means that the snow accumulation is now larger than the mass loss due to the outflow across the grounding line. The net mass gain thus causes the ice sheet grounding line to advance back to its original location, implying that this location is stable to small perturbations.

Bottom topography observations show that many ice streams in Greenland and Antarctica can find their grounding line on a retrograde slope after a modest retreat. This, together with the MISI mechanism, implies the possibility of a tipping point expressed as a catastrophic ice sheet collapse in a warmer climate. Major ice streams that lie on or near a retrograde bed slope and can therefore undergo an abrupt transition due to MISI include the Thwaites Glacier and Pine Island Glacier in west Antarctica, with a potential for significant sea level consequences. It should be noted, though, that several stabilizing feedbacks have been identified that can prevent ice sheets on retrograde slopes from collapsing, such that the expected effects of MISI are still uncertain.

10.2.5 Basal hydrology

Water at the base of an ice sheet can lead to a lubrication of the base, and therefore to ice flow acceleration, and may lead to the disintegration of an ice sheet. Some of the multiple ways for liquid water to accumulate at the base of the ice are considered next.

Moulins

These are well-like shafts going from the ice sheet surface all the way to the base, allowing surface melt water to flow to the ice base hundreds to thousands of meters below, lubricating the base, increasing the horizontal ice flow and there-fore increasing ablation. Consider the question of why water flowing down a moulin does not freeze on its way. The potential energy of a water parcel at the ice surface is converted to kinetic energy as it flows down, which is then con-verted to heat by friction with the moulin walls. To calculate the heating, assume the temperature increase applies to the falling water parcel itself, and write the energy conservation statement for a unit mass of water, stating that the inter-nal energy gain due to temperature increase equals the potential energy loss as $c_{p,water} \Delta T = g\,1000H$, where H is the height loss in kilometers. The temperature change is therefore

$$\Delta T = \frac{g\,1000H}{c_{p,water}} = \frac{9.8H}{4.2} = 2.33\frac{K}{km}\,H. \tag{10.2}$$

A drop corresponding to $H = 1$ km would generate heat that is enough to warm the falling water mass by over 2K. In reality some of this heat is transferred to the moulin walls, causing melting and preventing their closing in. In the context of global warming, more melt ponds at the surface in response to an atmospheric warming can potentially lead to more water transported from the ice surface to the base via moulins, potentially leading to ice flow acceleration.

Geothermal heat flux and basal melting

A heat flux of 0.05–0.1 W/m^2 on average flows out of the Earth surface as a result of the gradual cooling of the Earth core and is referred to as the geothermal heat flux. While this is a very weak heat flux relative to other fluxes in the climate system (air-sea heat fluxes are measured in hundreds of W/m^2), ice is a good heat insulator, and it can significantly slow the diffusion of geothermal heat from the base of an ice sheet toward the ice surface. This causes the heat to accumu-late near the ice base, possibly leading to melting, basal lubrication, and ice flow acceleration.

To calculate whether basal melting is expected due to the geothermal heat flux, and at what rate, consider the heat budget of the base of the ice. Let the incoming basal geothermal heat flux be denoted G, in units of W/m^2. The geothermal heat flux warms the ice base, while the ice sheet surface (at $z = H$) is typically cold, due to its high elevation and the atmospheric lapse rate. As a result, heat diffuses from the ice base to its surface, at a rate $-\kappa\,dT/dz$, where κ

is the conductivity of heat in ice (2.2 W/m/K), z is the upward distance such that $z = 0$ at the ice base, and $T(z)$ is the profile of temperature within the ice. In a steady state with no basal melting, the geothermal heat flux is carried upward through diffusion in the ice, so we can write

$$-\kappa \frac{dT}{dz} = G, \qquad \text{no basal melting.} \qquad (10.3)$$

This implies that the temperature profile within the ice is linear, $T(z) = a - Gz/\kappa$. To calculate the constant a, assume that the ice surface temperature, $T(z = H) = T_s$, is known, as it can be calculated from the atmospheric temperature or using a surface energy balance involving the incoming shortwave radiation and outgoing longwave radiation. The solution for the temperature profile within the ice is then

$$T(z) = T_s + \frac{G}{\kappa}(H - z),$$

and the temperature at the base $(z = 0)$ is therefore

$$T(0) = T_s + \frac{GH}{\kappa}. \qquad (10.4)$$

If the resulting basal temperature $T(0)$ is larger than the freezing temperature, we expect the base to be melting. If the base is melting, the basal temperature is equal to the melting temperature of ice, T_m, and the heat budget of the ice base needs to include the heat that goes into melting per unit area, M (W/m^2). The heat budget now balances the heat escaping upward from the ice base by diffusion with the geothermal heat arriving from below minus the heat that goes into melting,

$$-\kappa \frac{dT}{dz} = -\kappa \frac{T_s - T_m}{H} = G - M, \qquad \text{with basal melting.} \qquad (10.5)$$

We used the fact that in a steady state the temperature profile within the ice is linear to calculate the vertical temperature gradient from the top and bottom ice temperatures. The melting rate in $\text{kg s}^{-1}\text{m}^{-2}$ is then calculated using the latent heat of melting/freezing of ice, L_f $(3.36 \times 10^5 \text{ J/kg})$, as M/L_f. As a final case to consider, if melt water exists at the base, and if the temperature calculated via the solution to equation (10.3) is colder than the freezing temperature, then we deduce that basal freezing is expected. The latent heat of freezing released at the base is then represented by a negative M in equation (10.5) and can be calculated from that equation.

To summarize, in order to find out if basal melting occurs and its rate, we first calculate a basal temperature using the heat budget assuming no melting, (eqn 10.3). If the basal temperature solution given by equation (10.4) is above the freezing temperature, then we deduce that melting occurs, set the basal temperature to the melting point, and use equation (10.5) to calculate the melting rate at the base.

Note that the above solutions for the basal temperature and the basal melting rate depend on the upper ice surface temperature T_s. If the ice surface temperature increases due to a climate warming, this eventually affects the temperature at the base and can trigger melting or increase the melt rate there. However, the timescale by which surface temperature affects the ice base a couple of kilometers below the surface could be thousands of years because of the slow diffusion of heat in ice. This timescale may be crudely estimated from the thickness of the ice sheet H (m) and the diffusivity K (m^2/s) of heat in ice, which is related to the above conductivity as $K = \kappa/(\rho c_{p,ice}) \approx 1 \times 10^{-6}$ m^2/s, where $\rho = 900$ kg/m^3 is the ice density and $c_{p,ice} = 2100$ J kg^{-1}K^{-1} its specific heat capacity. To estimate the timescale by which surface processes can affect the base from the relevant physical parameters H and K, use dimensional analysis. The dimension of depth is meters, while that of K is m^2/s. The time it takes for a change in surface temperature to diffuse to the ice bottom and affect melting there can thus be estimated as the ratio H^2/K, which is many thousands of years for ice sheets whose thickness is in kilometers. The direct connection from surface temperature and surface melting to ice base lubrication provided by moulins, for example, is therefore a far more effective way of leading from climate warming to ice flow acceleration.

The ice melting temperature at the base of an ice sheet depends on the pressure there, which in turn depends on the thickness of the ice sheet. At sea level pressure, the melting temperature is, of course, $T_m = 0\,^\circ$C. For each kilometer of thickness of an ice sheet, the pressure increase at the ice base means that the melting temperature decreases by about $1\,^\circ$C, leading to a non-negligible effect.

Climate Background Box 10.1

Ice ages

Over the past 800,000 yr, Earth has gone through some eight glacial cycles, with ice ages separated by interglacial periods similar to the one that has existed over the past ~12 kyr (the Holocene); each glacial cycle has lasted about 80–120 kyr. During the last glacial maximum (LGM, 21 kyr before present [BP]), sea level dropped by about 130 m due to ice buildup over high-latitude continents, global average temperature was some 4 °C lower than at present, and atmospheric CO_2 dropped to 180 ppm, a significant reduction relative to the 280 ppm concentration during interglacial periods (green line in panel a in the accompanying figure). The temperature (e.g., the Antarctic temperature estimated from ice core data shown by the blue line in panel a) and land ice volume records have a characteristic saw-tooth shape representing gradual cooling and ice buildup over 90 kyr followed by a faster warming and melting over 10 kyr. During the LGM, northern North America was covered by an ice sheet of thickness of up to 3–4 km that extended all the way to present-day New York City, and Northern Europe was similarly ice covered. Panel b shows a deep ocean sediment isotopic record ($\delta^{18}O$, reflecting the fraction of different oxygen isotopes in buried plankton shells), which is a measure of both ice volume over land and the deep ocean temperature. It shows that ice ages began 2.7 Myr ago following a gradual cooling since the previous warm climates (Box 4.1) as shorter-period 41 kyr cycles, changing to a larger amplitude and a longer period of about 100 kyr around 0.8 Myr.

Figuring out the causes of glacial cycles is a test of our understanding of climate dynamics. In spite of significant progress, decades of studies of ice ages—which are the largest climate variability signal over the past million years—have still not yielded a consensus on the mechanism of the cycles, the reason for their timescale or their change of period at 0.8 Myr, or the reason for CO_2 changes. Successfully modeling the detailed climate of the LGM would increase our confidence in climate models that are used for future projections. A precise knowledge of the state of the climate system during ice ages also has consequences for present-day sea level studies, because the Earth crust is still rebounding from the weight of the LGM ice sheets, resulting in current regional sea level changes (as reviewed in section 4.2.2).

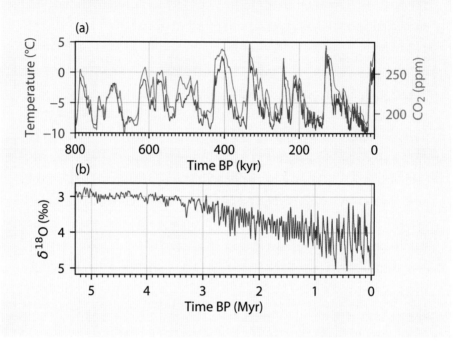

10.3 OBSERVED TRENDS AND PROJECTIONS

The ice mass over Greenland and Antarctica has been observed using the remarkable GRACE twin-satellite mission and its follow-on since 2002, quantifying ice mass changes via their small effects on temporal variations in the Earth gravity above the ice sheets. Figure 10.5a shows the cumulative mass change over this period, showing that Greenland lost some 4000 Gt during 2002–2020, or just over 200 Gt per year. This corresponds to about half a millimeter change in global mean sea level per year (out of a current GMSL rise rate of 3.5 mm/yr; see chapter 4). The observed change in Antarctica is an ice mass reduction of about 1500 Gt over the same period (Figure 10.5b). The typical timescales of ice sheets are hundreds and even thousands of years, and their mass may show long-term naturally occurring oscillations on such long timescales. Given this long-term

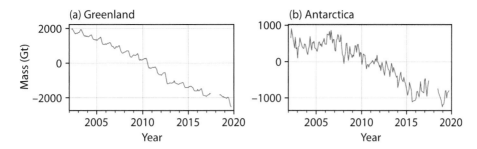

Figure 10.5: Observed changes in Greenland and Antarctica.
Time series of cumulative ice mass change for the Greenland and Antarctic ice sheets from the GRACE mission. The vertical axis is shifted arbitrarily so that only deduced differences in time are meaningful.

natural variability, it is very challenging to detect a signal due to anthropogenic climate change and separate it from the effects of natural variability for both Greenland and Antarctica, as this requires long records that are not available for these ice sheets. This challenge is similar to that of identifying a climate change signal in Arctic sea ice using long model runs (section 9.2) or in California droughts using long tree-ring records (section 12.3). In this case, the long timescales of ice sheets mean that much longer records are required for a confident detection of an ACC signal.

Proceeding to future projections, we first consider SMB changes for Antarctica and Greenland as predicted by a climate model. Figure 10.6a shows a map of projected net SMB changes over Antarctica, while panel b shows it for Greenland. Consistent with the colder temperature regime in the schematic Figure 10.2, Antarctica shows a projected net increase in accumulation, especially over its margins. Greenland shows a more mixed signal, with net surface melting/sublimation in the west and a surface mass gain in the east. In these calculations, it is assumed that only snow accumulation contributes to the SMB, while rain just runs off the ice sheets. This affects mostly the SMB estimate for Greenland, where although the total precipitation is projected to increase in a warmer climate (not shown), a significant portion of the increase in precipitation is in the form of rain rather than snow. This assumption that only snow accumulation contributes to the SMB is not easily justified, as it is difficult to estimate what portion of the liquid precipitation will indeed contribute to the runoff rather than freeze and contribute to the SMB. In any case, the projected

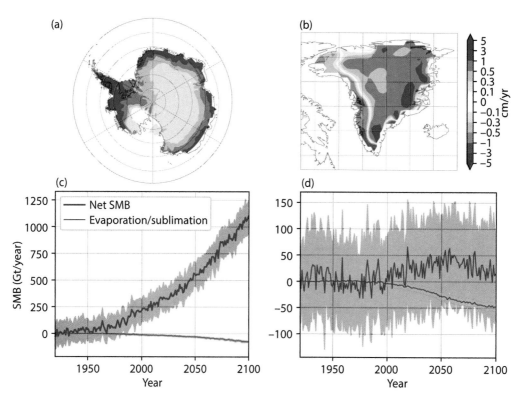

(a)   (b)

(c)   (d)

Figure 10.6: RCP8.5 projections of surface mass balance changes over Greenland and Antarctica.

(a) A contour plot of net SMB change from 1920 to 2100 in cm/yr, averaged over 30 climate model ensemble members. The net change is calculated as the change in snow accumulation minus the change in sublimation rate, calculated as the surface latent heat flux divided by the latent heat of sublimation. The blue shades indicate a gain in SMB over Antarctica. (b) Same, for Greenland, where a reduction in SMB is projected over the margins, especially in the west. (c) The blue line shows a time series of net surface mass balance (snow accumulation minus sublimation) for the AIS in Gt/yr, averaged over 30 climate model ensemble members. The light-blue shading indicates one std over the ensemble members. The red line and shading show the contribution of sublimation, calculated as in a. (d) Same, for Greenland.

addition of ice mass to Antarctica in a somewhat warmer climate suggests that the observed mass loss in Figure 10.5b may be a transient trend due to natural variability rather than a response to global warming.

These climate model projections of the SMB of Greenland and Antarctica involve significant uncertainties, both due to the issue of runoff indicated above and due to the fact that precipitation is generally one of the climate variables most prone to model uncertainties, as discussed in chapter 7 and in section 12.6. Yet an even larger uncertainty is associated with attempting to model what is often referred to as the *dynamical* or *fast response* of the ice sheets, involving ice stream acceleration or ice sheet collapse, for example. This is due to the difficulties of correctly representing many of the processes reviewed above in an ice flow model, from basal hydrology to MISI to calving. Recent studies show that future projections of ice sheet mass on the scale of hundreds of years are most sensitive to these sub-grid-scale parameterized processes; thus projections regarding the fate of these ice sheets on timescales beyond 2100 still involve great uncertainty.

10.4 WORKSHOP

A Jupyter notebook with the workshop and corresponding data file are available; see https://press.princeton.edu/global-warming-science.

1. **Observations and projections:**

    (a) Plot the time series of observed cumulative mass changes in Greenland and Antarctica from the GRACE mission.

    (b) Contour the projected net SMB change under the RCP8.5 scenario from 1920 to 2100 for Greenland and for Antarctica. Plot time series of total net SMB over Greenland and over Antarctica for the same scenario. Explain your plots, discussing and rationalizing both the spatial and temporal structure of the projected changes.

2. **Ice stream acceleration:** By what factor should ice streams in Greenland accelerate now in order to contribute a 1.5 m sea level rise by 2100?

3. **Positive degree days:**

    (a) Calculate and plot PDD maps for Greenland and Antarctica for the present-day and for year 2100, based on a temperature projection from the RCP8.5 scenario.

    (b) Estimate the expected percent increase in melting rate (assuming melting is proportional to PDD) and the consequences for sea level rise.

    (c) Compare your SMB results for Greenland and Antarctica to the above calculated required acceleration of ice streams, and speculate which mechanism is likely to contribute a large sea level increase by 2100.

4. **Calving:**

    (a) Buoyancy forcing: What is the maximum protruding section length L in Figure 10.3b that can be stable to buoyancy forces?

    (b) Hydrofracturing and buoyancy forcing: Consider a section of grounded ice of height $H = 2$ km, bordering with the ocean such that the top of the ice is above sea level and the ice in question extends a distance $L = 50$ m from the ocean. Calculate the minimum depth of water that will lead to the flotation and therefore breakup of the ice, taking into account the adhesion to the rest of the ice via the yield stress τ_c. How would your answer change if there was a hydrofracture extending to half the thickness of the ice?

(c) Cliff instability: What is the highest cliff height H in Figure 10.3c that allows the marked 45° section to be stable?

5. **Basal hydrology:**

   (a) Basal temperature and melting: Calculate the basal temperature, and where relevant also the melt rate in millimeters of melt water produced per unit area per year, given a surface ice temperature of $T_s = -40\,°C$, geothermal heat flux of $G = 0.07\ W/m^2$, and ice thicknesses of $H = 1$, 2, and 3 km.

   (b) *Optional extra credit:* **Moulins:** Consider a 2.5 km deep, 1 m radius moulin in Greenland, and suppose all the heat due to the potential energy of the melt water flowing down the moulin is used to widen the moulin. What water flow, in m^3/sec, is required to enlarge the radius to 2 m during the one month of peak melt season?

6. **Guiding questions to be addressed in your report:**

   (a) What are the climate and socioeconomic effects of a possible Greenland ice sheet melt in a warmer future climate?

   (b) List, explain, and compare different processes that may lead to future changes to the ice sheets in Greenland and Antarctica.

   (c) Discuss processes that may lead to a *rapid* ice sheet collapse/ retreat.

   (d) Discuss factors that may make this retreat irreversible on a timescale of hundreds and possibly thousands of years. Which portions of which ice sheets may be affected?

   (e) Suggest what elements of Greenland and Antarctica should be monitored to identify early warning signals.

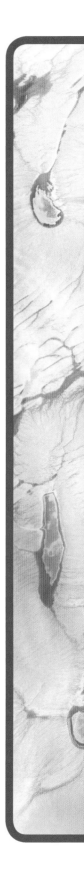

MOUNTAIN GLACIERS

Key concepts

- Observed retreat over the past 150 yr, acceleration in recent decades
- Surface mass balance, equilibrium line, accumulation and ablation zones
- Glaciers as climate proxies: ice cores and glacier length records
- Glacier ice flow, retreat due to warming and changes in surface mass balance
- Retreat due to exit from Little Ice Age versus anthropogenic climate change

The retreat of mountain glaciers since the 19th century and the corresponding before and after pictures from prominent glaciers worldwide are iconic symbols of global warming. The retreat is almost universally seen in mountain glaciers at high latitudes and low latitudes, and over all continents with mountain glaciers, and there are many important and interesting lessons to be learned from this widespread retreat. Mountain glaciers, especially in low latitudes, provide runoff water to large populations, complementing weaker rainfall during dry seasons and providing needed water during years of drought. Their shrinking area therefore has significant socioeconomic effects, as well as effects on the surrounding natural habitats. The melting of mountain glaciers has been estimated to contribute about 20% of global mean sea level rise since year 2000 and is expected to lead to further sea level rise in the future. Figure 11.1 shows multiple records of glacier length (the location of the front of the glaciers relative to their location in 1960); it is evidently clear that a strong retreat is occurring for nearly all glaciers, as will be further analyzed below. Because the retreat seems to have started around 1850, prior to a significant increase in greenhouse forcing, one needs to address whether it is a result of the exit from the naturally occurring cold period known as the *Little Ice Age* (extending roughly from 1300 to about 1850) or of anthropogenic climate change. We address this below in several ways.

A mountain glacier is maintained at a steady state by snow accumulation at its upper range, known as the accumulation zone, balanced by ice flow to lower and typically warmer altitudes, where ice melts within the ablation zone (as

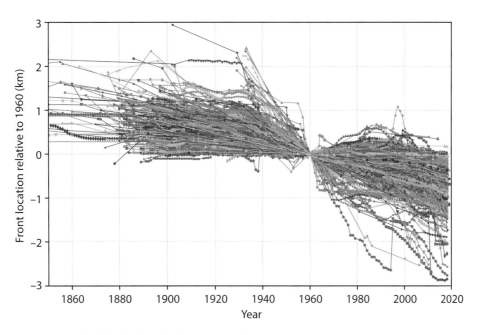

Figure 11.1: A global retreat of mountain glaciers.
Glacier length time series for 633 records, relative to their position in 1960.

reviewed in the context of the Greenland ice sheet in section 10.1). The equilibrium line altitude (ELA) is the altitude separating the accumulation and ablation zones. The accumulation minus melting and sublimation at the ice surface is referred to as the surface mass balance (SMB). Melting is often well correlated with the positive degree day (PDD) measure (section 10.2.2), and the surface mass balance of mountain glaciers is strongly affected by the atmospheric temperature. Studies of many glaciers show that, generally, a mere 1 K of atmospheric warming is equivalent to a decrease of 25% in the accumulation rate.

Some of the most charismatic mountain glaciers, whose retreat seems to be accelerating in recent decades, are found in the tropics, be it in Africa (Kilimanjaro, Mount Kenya), the Andes, or other tropical locations. It is worth noting that because of the tropical lapse rate feedback (section 3.2), a surface warming in the tropics leads to a larger warming at higher altitudes, making tropical glaciers especially vulnerable to climate warming. While changes to clouds (which control the SW radiation reaching the ice surface and enhancing melting) and a changing rate of accumulation may play a role in these observed changes, a careful analysis suggests that warming is likely a dominant factor in the retreat of these glaciers.

We start with a quantitative analysis of the retreat (section 11.1). Mountain glaciers contain several indicators of climate change history, as discussed in section 11.2: First, the detailed history of glacier length may be quantitatively related to the warming responsible for the retreat. Second, ice cores drilled over the past decades in high-altitude mountain glaciers, in a heroic effort of a few climate scientists, contain a detailed climate record. Finally, we discuss the dynamics of glacier flow and how glaciers are expected to respond to climate warming (section 11.3). A brief final perspective of the retreat of mountain glaciers, and in particular a summary of why the accelerated retreat in recent decades is very likely related to anthropogenic climate change, is given in section 11.4.

11.1 OBSERVED RETREAT

To further quantify the overall glacier retreat in Figure 11.1, consider the 3 yr bin-average of the front glacier locations in Figure 11.2a, with the number of observations per bin in Figure 11.2b. Interesting features in this averaged glacier length record include an accelerating decline since 1900, an even stronger recent acceleration since 1990, and a pause in decline around 1970 that is reminiscent of the global warming "hiatus" that occurred around the middle of the 20th century (section 3.3) and whose relation to the glacier trends will be further

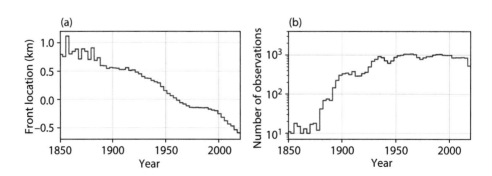

Figure 11.2: Averaged glacier length trends.
(a) A 3 yr bin-average of the glacier length records in Figure 11.1. (b) The number of observations per bin.

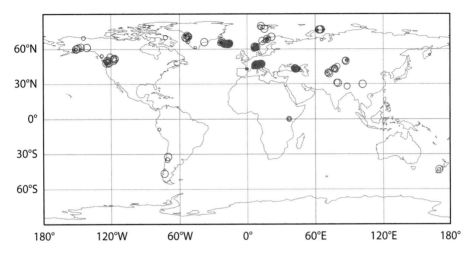

Figure 11.3: Global distribution of trends.
Glaciers with observed length time series in the world glacier monitoring service data base, marking the 622 glaciers with negative trends in red and the 11 with positive trends in blue. The symbol sizes represent the magnitude of the trend. The blue symbols are drawn on top of the red ones to ensure that the few locations with positive trends are clearly highlighted. (The data set used to prepare this figure does not include all mountain glaciers.)

analyzed below. Fitting a linear trend to each record of glacier length, we find that the vast majority of glaciers (622 of them) show a negative trend (reducing length), and only 11 show an increasing length trend; the locations of positive and negative trends are shown in Figure 11.3. The retreat is seen to be occurring worldwide, and even the few glaciers that seem to show a positive trend are typically embedded in large groups of retreating glaciers. The few increasing length trends in individual glaciers can be the result of local changes to the spatial patterns or to the seasonality of accumulation that more than compensate for the increase in temperature. A glacier can also increase in thickness following a naturally occurring *surge* event that previously led to thinning and mass loss.

Given the dramatic retreat of these glaciers, and in particular their acceleration in the recent decades since 1990 (Figure 11.2a), one might wonder if it is still possible for this to be a delayed response to the naturally occurring warming at the end of the Little Ice Age over 150 yr ago. Very powerful arguments against this possibility are provided by several lines of evidence, and we begin with carbon dating of exposed plant material under retreating glaciers. These plants are found well preserved and can even regrow once exposed, indicating that they were preserved frozen continuously since the glacier expansion that buried them. There is a rich record of such plant material that is collected immediately when it is exposed under retreating glaciers worldwide. The carbon dates of these plants thus represent the time during which the plant grew, and therefore the last time the area in question was exposed and glacier-free. A date of 1500 years, say, implies that the glacier expanded into this area that long ago and never retreated from it since. Sufficiently old dates thus rule out the possibility that the mountain glaciers we see retreating today have developed only since 1350 or so, when the Little Ice Age started, and are merely retreating now back to their natural location. There are numerous examples of such last-exposure dates, including 5000–5300 yr BP for the Quelccaya ice cap in Peru, 40,000 yr for the Baffin Island, 3400–4000 yr for Svalbard, and more. These are all in contrast to some carbon dates from the Ellesmere Island, where rapid glacier retreat is exposing terrain for the first time only since the Little Ice Age. Overall, it is clear that numerous glaciers are retreating to the dramatic currently observed extent for the first time in thousands of years, long before the onset of the Little Ice Age, making the present retreat exceptional and unlikely to be a signal of natural climate variability.

11.2 MOUNTAIN GLACIERS AS A CLIMATE INDICATOR

We now consider two ways in which information obtained from mountain glaciers can provide us with clues on temperature changes: first, glacier length is strongly correlated with temperature; and second, ice cores taken from mountain glaciers contain a rich archive of past climate changes.

11.2.1 Reconstructing temperature from glacier extent

The archive of mountain glacier length records for hundreds of glaciers over the world, some of which are hundreds of years long (Figures 11.1, 11.2, and 11.3), is a rich source of climate information. Because, as mentioned above, glacier extent is very sensitive to temperature, one is tempted to use the glacier extent to reconstruct temperature changes over the past century and a half or so.

In addition to the above-mentioned observation that glacier length seems more sensitive to temperature than to a change in accumulation, we note that the spatial patterns of observed precipitation changes tend to be of a smaller scale than that of observed warming. These two observations, plus the fact that glacier retreat is an essentially global phenomenon (Figure 11.3), suggest that it may be possible to relate glacier length to temperature alone. The adjustment time of glaciers to temperature changes can vary from decades to hundreds of years, depending on the topographic slope, accumulation rate, and more. Let us first consider for simplicity the unrealistic scenario that a given glacier is in equilibrium with the local temperature because the temperature is changing very slowly, allowing the glacier length to fully adjust to the temperature at any given time. In that case, we can relate the glacier length anomaly to the temperature anomaly. Let the glacier length anomaly, defined as its deviation from the length in a specified reference year, be L', and let the local temperature anomaly relative to that in the reference year be T'. Assuming a simple linear relation between length and temperature, one writes $L' = -cT'$. The proportionality constant is the *climate sensitivity* of the glacier, measured in km/K.

However, over the past century and a half, temperature has changed too quickly to allow glacier lengths to equilibrate at any given time, and this equilibrium relation cannot be assumed to hold. Instead, the glacier length continuously adjusts toward its equilibrium with the changing atmospheric temperature, with a typical timescale τ (typically decades to hundreds of years; see below). The adjustment may be described by the simple differential equation

$$\frac{dL'(t)}{dt} = -\frac{1}{\tau}\left(L'(t) + cT'(t)\right). \tag{11.1}$$

We can intuitively understand this equation by assuming a simple scenario: Suppose that at $t = 0$ the glacier had a length L_0' when the temperature has changed to some new fixed value T_0' and remained there, and that the glacier length is now adjusting to the new temperature. The solution to equation (11.1) for this case of a constant temperature is

$$L'(t) = (L_0' + cT_0')e^{-t/\tau} - cT_0', \tag{11.2}$$

which may be verified by noting that it satisfies both the differential equation and the prescribed initial condition for the glacier length at $t = 0$. We see from this solution that as the first term decays, the glacier length is approaching its

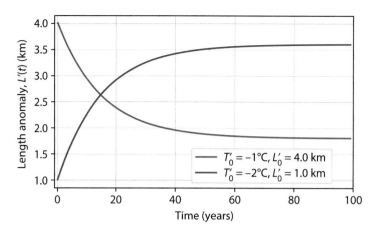

Figure 11.4: Glacier adjustment to temperature perturbation.
Two idealized adjustment scenarios of glacier length based on solution (11.2), assuming a glacier adjustment time of $\tau = 15$ yr, and based on the initial lengths and perturbation temperatures indicated.

final equilibrium state $-cT_0'$ exponentially with a timescale τ. Figure 11.4 shows two example scenarios, one in which the glacier is advancing and one in which it is retreating, both showing the exponential approach to the new equilibrium. More generally, we can now understand equation (11.1) to describe the exponential approach of the glacier length $L'(t)$ toward its equilibrium value with the continuously changing temperature perturbation $T'(t)$, with an adjustment time τ.

We next proceed to the analysis of the more realistic case of a temperature that keeps changing, in order to be able to relate observations of global mean temperature and glacier length records over the past century and a half. For this purpose, we take $L'(t)$ in (11.1) to be the glacier length anomaly relative to that at year 1960 and $T'(t)$ to be the temperature anomaly relative to that year. Equation (11.1) may be rewritten as an equation for the temperature,

$$T'(t) = -\frac{1}{c}\left(L'(t) + \tau\frac{dL'(t)}{dt}\right). \tag{11.3}$$

We can evaluate the length anomaly and its time derivative, which both appear on the RHS, from the mean glacier anomaly extent in Figure 11.2a. We take the adjustment timescale τ and the climate sensitivity c as unknowns, although for

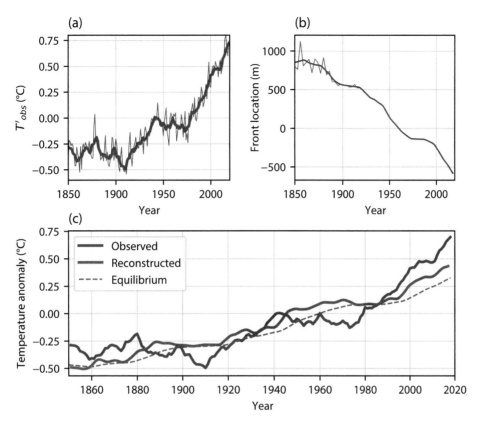

Figure 11.5: Relating temperature to glacier length.
(a) Globally and annually averaged surface temperature (T'_{obs}, red) and its smoothed version used for the analysis of glacier length and global temperature (blue). (b) The bin-averaged glacier length from Figure 11.2a, interpolated to 1 yr resolution (red) and smoothed (blue). (c) The optimal solution for the reconstructed global mean surface temperature calculated from the smoothed glacier extent time series using eqn (11.3) (red) together with the observed smoothed temperature redrawn from panel a (blue) and the reconstructed equilibrium temperature (dashed red).

well-studied glaciers they may be calculated from local topographic and meteorological conditions. We then use the observed global mean surface temperature relative to 1960, denoted $T'_{obs}(t)$, to calculate the glacier climate sensitivity and adjustment timescale that minimize the difference between the temperature reconstructed from glacier length and observed temperature anomalies. That is, we find c and τ that minimize $\int_{1850}^{2020} \left(T'(t) - T'_{obs}(t) \right)^2 dt$. This optimization is done using the smoothed mean glacier length and globally mean surface temperature in Figures 11.5a,b. The resulting reconstructed temperature is shown by the solid red line in Figure 11.5c. It seems to capture the main features of the observed temperature shown by the solid blue curve, in particular the overall increasing trend, recent acceleration in warming, and—remarkably—the "hiatus" periods during 1940–1980 and during the first decade of the 21st century. The optimization yields a climate sensitivity value of $c = 1.8$ km/K and $\tau = 11.2$ years. The dashed red line in Figure 11.5c shows the equilibrium temperature calculated from the glacier length as $T' = -L'/c$. This temperature is a significantly less good fit to the observed record, strengthening the case that glaciers respond to temperature changes with some delay as represented by the timescale τ.

This calculation is very heuristic: Climate sensitivity is expected to vary dramatically from glacier to glacier and with latitude, and the adjustment time strongly depends on the slope on which the glacier lies, as well as on local meteorological conditions. In addition, the geographic distribution of glacier records is not uniform and not of global coverage, and the average glacier length therefore cannot be expected to respond to the *global mean* surface temperature. These results should thus be seen as very qualitative. In particular, the glacier response time found here is an underestimate for many glaciers. In a recent study, detailed information about c and τ for 169 glaciers was used to reconstruct the global temperature with a remarkable success, demonstrating the relevance of the simple relation between glacier length and temperature in equation (11.1). This success is another evidence that the global retreat of mountain glaciers over the past few decades is a response to anthropogenic global warming, as opposed to being due to the exit from the Little Ice Age that terminated in the 19th century.

11.2.2 Ice cores from mountain glaciers

Numerous ice cores have been drilled over the past decades in mountain glaciers, to depths of up to hundreds of meters. These cores show annual bands of insoluble dust that can be used to date the ice layers. The isotopic composition (in particular $\delta^{18}O$ within the ice) is generally related to climate factors and in many cases may serve as a reasonable proxy for temperature. Figure 11.6b shows a decadal bin-average of annual-resolution isotopic record from the Quelccaya Summit Ice Dome core in the Andes, Peru (latitude 13°S). This long record, heroically obtained multiple times over the past decades, suggests that

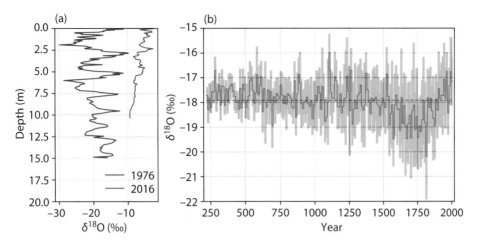

Figure 11.6: Isotopic records from the Quelccaya ice cap in the Andes, Peru (latitude 13°S).

(a) Two high-resolution shallow ice cores showing the presence of a seasonal cycle in 1976 (blue) and its elimination by surface melting and percolation of melt water by the time the 2016 core was drilled (red, shifted by 10‰ to improve legibility). (b) A decadal bin-average of a long isotopic record from the Quelccaya Summit Ice core. The cyan shading indicates plus and minus 1 std for each decade.

the trend over the last few decades is very unusual relative to the past 2000 years, consistent with the (shorter) glacier length data discussed above.

In addition, a higher resolution core drilled in 1976, shown in blue in Figure 11.6a, clearly shows about seven seasonal cycles. However, a core drilled nearby in 2016 (red, shifted by 10‰ to the right for visibility) hardly shows a seasonal cycle signal. It turns out that melting at the surface, and percolation of melt water into the ice, smoothed out the seasonal cycle. The examination of longer records from that area suggests that such surface melting events are unusual during the past many hundreds to thousands of years and are therefore likely related to the observed anthropogenic warming in recent decades.

11.3 GLACIER DYNAMICS

In order to obtain further insight into the response of mountain glaciers to climate change, we use a simulation of a glacier over an idealized mountain-like topography, forced by a prescribed idealized SMB. Figure 11.7a shows SMB profiles before a presumed climate warming (solid blue line) and after the warming

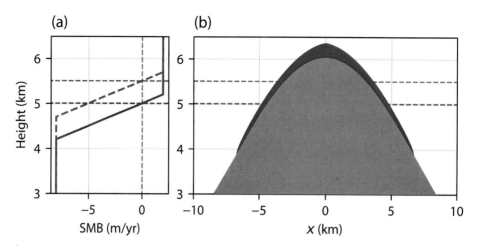

Figure 11.7: A simulated mountain glacier before and after a prescribed warming scenario.
(a) Surface mass balance for two scenarios (solid blue vs dashed red), showing also the corresponding equilibrium line altitudes (horizontal dashed lines). (b) The steady solutions of a shallow ice approximation model for glacier height for the two scenarios.

(dashed red). Note that the ELA (where the SMB crosses the zero value and is marked by horizontal blue and red dashed lines) increases in this scenario from 5 km to 5.5 km. Assuming the ELA depends on the atmospheric temperature, and with a lapse rate of, say, 6.5 K/km, such an increase of the ELA corresponds to a warming of just above three degrees Kelvin. The ice thickness solution for the first SMB is in blue in Figure 11.7b. The figure shows that much of the glacier exists below the ELA, as far down as less than 4 km high, more than 1 km below the ELA, where the SMB is strongly negative (net melting of about 8 m/yr). The glacier extension below the ELA is supplied by ice flow from the accumulation zone above the ELA. The ice flow is driven by gravity pulling the glacier down the slope. When the ELA rises to 5.5 km due to climate warming, the glacier retreats to the solution in red in Figure 11.7b, and the lowest glacier altitude rises from 3.9 km to about 4.7 km, reflecting a shortening of the glacier length, as expected. Again, the lowest portion of the glacier lies below the new ELA, being fed by ice flow from the now smaller accumulation zone above the ELA.

Figure 11.8 shows the result of a scenario in which the glacier was at a steady state with the lower prescribed ELA, which was then abruptly raised. The glacier thickness as a function of horizontal location is in Figure 11.8a, with the

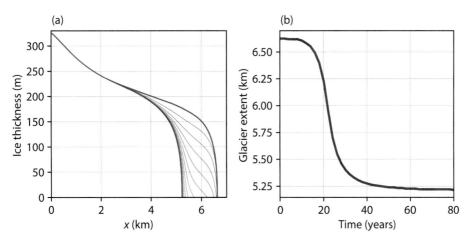

Figure 11.8: Time-dependent glacier adjustment to change in SMB.
The time-dependent transition between the blue and red solutions in Figure 11.7b. (a) Glacier thickness as a function of horizontal distance every 2 yr after the ELA changed, with progressing times shown by blue to red thin lines. (b) Glacier length as a function of time during the transition.

thickness every 2 yr indicated by the thin lines with colors changing from blue to red as the simulation progresses. Figure 11.8b shows that the glacier length (measured along the horizontal x-axis of Figure 11.7b rather than down the mountain slope) takes some 10 yr to start responding but then changes rapidly starting at year 20, followed by a slower equilibration of the glacier length to the new ELA position. The glacier response to climate warming may now be understood as follows. The warming changes the SMB, raising the ELA. This leads to a melting and thinning of the lower part of the glacier as seen by the first few thin lines of Figure 11.8a. Initially the thinning is not accompanied by retreat of the glacier front (the first few thin lines terminate at the original location). The reduction in the glacier thickness in the lower part of the glacier strengthens the downslope surface gradients of surface ice height. This then results in an initially enhanced ice flow downhill, accompanied by stronger melting in the lower parts of the glaciers and by a retreat of the glacier front, until the new equilibrium is reached.

The slow final states of the equilibration seen in this simulation (years 40–80 in Figure 11.8b) are in marked contrast with the observed global glacier retreat which seems to have actually accelerated in the past 20–30 yr (Figure 11.2a).

This again suggests that this observed recent accelerated global glacier retreat is unlikely to be a response to the termination of the Little Ice Age in the 19th century, as we should have been in the slow final adjustment to that change by now.

The results in Figures 11.7 and 11.8 were obtained by solving what is known as the *shallow ice approximation* (SIA) for glacier or ice sheet flow, based on the reasonable assumption that the ice thickness is significantly smaller than its horizontal extent.

11.4 MOUNTAIN GLACIER RETREAT IN PERSPECTIVE

The above results suggest that, unlike some other issues surrounding global warming, there is little uncertainty surrounding the existence of an already observed global mountain glacier retreat and its being a result of the robustly observed warming global temperatures. Figure 11.3 leaves very little doubt that this is a global event. And we have discussed multiple lines of evidence that the accelerated retreat in the past few decades (different from the initial retreat during the 19th century, and perhaps even from that during the first half of the 20th century) is not likely a result of an exit from the naturally occurring Little Ice Age that lasted from 1300 to about 1850. To reiterate, the evidence is as follows:

- Carbon-dated exposed plant material suggests the last time such a retreat of mountain glaciers occurred was many hundreds to thousands of years ago, long before the start of the Little Ice Age.
- The close relation between glacier length records and global temperature indicates that the glacier retreat during the past few decades must be as unusual as the recent warming seen in the temperature record from independent sources.
- The accelerating rate of retreat seen in recent decades is in contrast to expected slowdown in the response to the termination of the Little Ice Age over 150 yr ago.
- The adjustment time calculated for most glaciers is short (one to a few decades), suggesting that any adjustment to the exit from the Little Ice Age is long over.

- Isotopic composition in mountain glacier ice cores considered as a temperature proxy suggests that the 20th-century warming is unusual within the past 1500 yr or so.
- Surface melting events observed in tropical ice cores in the 21st century seem to have not occurred in the previous many hundreds to thousands of years.

With such diverse evidence, one can conclude fairly confidently that the accelerated global retreat of mountain glaciers in the past few decades is a result of the anthropogenic global warming over this period and that it is likely to continue at a rate that may eliminate many of these glaciers in the coming decades. In other words, the global trends seen in Figures 11.1 and 11.3, especially since 1990 or so, make it very obvious that a climate change signal has been detected. Furthermore, the above bullet points make a strong case that the trends can be attributed to a greenhouse gas–driven global warming as opposed to alternative possible explanations (e.g., exit from the Little Ice Age). The reader is advised to return to section 3.4.1 for a general discussion of detection versus attribution and to sections 9.2 and 12.3 for example statistical analyses that can be used to quantify the issue of detection and also applied to the problem of mountain glaciers.

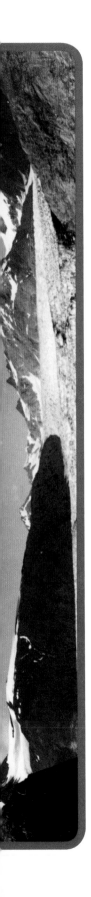

11.5 WORKSHOP

A Jupyter notebook with the workshop and corresponding data file are available; see https://press.princeton.edu/global-warming-science.

1. **Glacier length records:**

    (a) Plot the time evolution of all glacier lengths relative to 1960.

    (b) Plot the length time series of one glacier from each of the following: Argentina, Peru, New Zealand, United States, Switzerland, and Tajikistan. Choose the glacier with the record with the largest number of data points for each of these countries.

    (c) Calculate and plot the bin-average over all glaciers of glacier length relative to 1960 as a function of year.

2. **Temperature and glacier length:**

    (a) Plot glacier length over time for the Morteratsch Glacier (Vadret da Morteratsch) in Switzerland.

    (b) Plot the annually and globally averaged surface temperature.

    (c) Calculate and superimpose the temperature reconstructed from the length record of the above glacier using equation (11.1): Find, via trial and error, a glacier sensitivity c and a response time τ that lead to a temperature that is qualitatively similar to the globally averaged one, with the "hiatus" periods and accelerated warming in the second half of the 20th century.

    (d) Superimpose a line plot of the temperature that is in equilibrium with the glacier length at any given time using the sensitivity calculated above. Discuss the character and reasons for the difference between this equilibrium temperature and the one reconstructed using equation (11.1).

3. **Idealized glacier length adjustment scenarios:**

    (a) Plot time series of idealized glacier length adjustment scenarios with a temperature that changes abruptly at $t = 0$ and then remains constant, and using the exponential solution to equation (11.1). Use the same c and τ parameters calculated in question 2. For the first scenario, let the initially imposed temperature anomaly T_0' be $-1\,°C$ and the initial length L_0' be 4000 m. For the second scenario, $T_0' = -2\,°C$ and $L_0' = 1000$ m.

    (b) *Optional extra credit:* **Glacier length response to a varying temperature:** Find and plot the solution to equation (11.1) for a linearly changing temperature. Use start and end anomaly temperatures of $T_0 = -0.25$ K, $T_f = 1$ K, climate sensitivity of $c = 3000$ m/K, adjustment time of $\tau = 20$ yr, a total period of 140 yr, and an initial length anomaly of $L_0 = 6000$ m. Explain the structure of the solution you find.

4. **Isotopic records from Quelccaya ice cores:**

   (a) Plot the two provided individual ice cores from 1976 and 2016, and rationalize the amplitude of the seasonal cycle seen in them.

   (b) Plot the decadal bin-average of the longer isotopic record for the past 2000 years, add shading representing 1 std in each decade. Discuss the trends seen since 1750 and in the last few decades in the context of the longer observed record.

5. *Optional extra credit:* **The global picture:** Calculate a linear regression for each of the given glacier records from the global data set. Draw a global map with circles indicating the latitude and longitude of glaciers, with red indicating negative length trends (slope). Make the size of the circles reflect the magnitude of the trend. Superimpose blue crosses at the location of glaciers showing positive trends.

6. **Guiding questions to be addressed in your report:**

   (a) Why are mountain glaciers important?

   (b) How did they change over the past 200 yr?

   (c) What are possible reasons for changes seen in mountain glaciers over different periods during the past 200 yr?

   (d) What are the timescales of observed and expected changes?

   (e) Investigate and discuss projections for the next 20, 50, and 100 yr for different glaciated mountain regions at both low and high latitudes, not including Greenland and Antarctica.

   (f) Discuss the socioeconomic consequences of the projected further retreat of mountain glaciers, focusing on a couple of key regions as examples.

DROUGHTS AND PRECIPITATION

Key concepts

- Precipitation, evaporation, and soil moisture
- Droughts driven by remote SST changes due to natural variability modes such as El Niño or the Indian Ocean dipole
- Reconstructing past droughts, tree rings, and the detection of anthropogenic climate change
- Future projections: two case studies, the Sahel and the Southwest United States
- The expected expansion of the Hadley circulation and changes to the location of desert bands
- The "wet getting wetter, dry getting drier" global-scale precipitation projection
- Intensification of extreme precipitation events
- Bucket model for soil moisture

D roughts are prolonged dry periods with multiple possible significant impacts, including water shortages, crop failures, soil erosion, mass migrations, and other long-term environmental, economic, and health impacts. Droughts rank second (after tornadoes and hurricanes) among weather-related natural disasters leading to significant economic damages. Severe droughts affect over 50 million people per year. Importantly, droughts are defined relative to long-term mean conditions and therefore reflect deviations from such a mean rather than absolute precipitation or soil moisture levels. The precise thresholds of dryness or duration used for the definition of droughts vary from region to region, and typical durations vary from months to decades.

Droughts are categorized into four general types: 1) *meteorological or climatological,* 2) *hydrological,* 3) *agricultural,* and 4) *socioeconomic.* The first three types are defined by physical or biological parameters, while the fourth centers on the impacts on society. A meteorological or climatological drought is defined simply in terms of the magnitude and duration of a precipitation shortfall. Hydrological droughts occur due to a precipitation shortfall that depletes surface or subsurface water supplies. Agricultural droughts involve impacts on agriculture via a soil moisture deficit and decrease in surface runoff. Hydrological droughts often lag meteorological droughts but may precede agricultural droughts. Droughts in recent decades raised the concern that anthropogenic climate change may have already made such events more intense and/or more frequent. While one still cannot determine with confidence that droughts have increased on a global scale, their frequency and intensity may have increased by some estimates in the Mediterranean and West Africa and decreased in central North America and northwest Australia.

The mean annual precipitation is expected to either increase or decrease in different regions in a warmer climate; precipitation changes on a *regional* scale, while highly relevant and important to predict, are among the variables whose prediction is least consistent among different climate models. There are some expected *global*-scale precipitation trends, such as the idea that changes will follow the pattern of "wet getting wetter and dry getting dryer," although it has not been found to be robust on a regional scale. Similarly, a latitudinal migration of desert bands in a warmer climate due to changes in the atmospheric general circulation has been suggested, but this idea seems again not to allow robust regional predictions of mean annual or seasonal precipitation changes. The occurrence of heavy precipitation events (precipitation extremes) is expected to increase in a warmer climate. That is, more of the precipitation is expected to fall in a few, short extreme precipitation events. Possible damages from such an expected future intensification of extreme precipitation events include more frequent and stronger floods, land slides, crop losses, and drinkable water losses due to increased runoff as soil storage does not keep up with precipitation rate. Flooding is among the most costly natural hazards, with billions of dollars in damages each year, and an intensification of this risk factor is expected to have significant socioeconomic effects.

Our purposes in this chapter are (1) to understand what causes droughts on a regional scale, and why it is difficult to predict future changes to these events; (2) to survey a couple of examples that will demonstrate the uncertainty in future projections; (3) to understand what can be said about the expected changes to both mean precipitation at different regions and latitudes and to the occurrence of precipitation extremes.

We next define some basic terms (section 12.1), examine how droughts happen as a response to a remote persistent deviation of the sea surface temperature from its long-term mean (that is, SST anomaly; see section 12.2), and discuss how one can detect climate change in a drought record using a long reconstructed record of drought events and non-parametric statistics (section 12.3). Then, we examine how tree-ring data can be used to reconstruct such a long drought record, which can put recent droughts in perspective (section 12.4), and discuss future projections in two example regions (Sahel and the Southwest United States; see section 12.5). While drought events tend to be regional, some projected changes to the societally important distribution of precipitation are expected to be regional and some of a global scale, and we examine three perspectives of this issue in section 12.6. First, we discuss the projected poleward expansion of the atmospheric Hadley circulation, which may lead to shifts in the location of the desert belts (Sahara, Kalahari, and others; see section 12.6.1). Second, we attempt to understand why precipitation projections follow an interesting pattern of "wet getting wetter and dry getting drier" (section 12.6.2), although mostly over the ocean and less so over land, where projections are more uncertain. Third, we attempt to understand and quantify the projected increase in extreme precipitation events (section 12.6.3). Finally, we introduce and carefully analyze a *bucket model* for soil moisture, which explains how changes to precipitation affect soil moisture, a critical variable in the context of droughts and agriculture in particular (section 12.7).

12.1 RELEVANT PROCESSES AND TERMS

Soil moisture: The amount of water stored in the soil, quantified as explained in section 12.7.

Evapotranspiration: The sum of evaporation from the soil plus evaporation from stomata in leaves that is fed by transpiration, the movement of water within plants from roots to leaves.

Potential evapotranspiration (PET): This is the amount of evapotranspiration that would have occurred given the meteorological conditions (wind, air temperature, humidity, insolation) if the surface water source were not a limiting factor. PET therefore measures the "demand" for evaporation as determined by the meteorological conditions.

Palmer Drought Severity Index: The PDSI was originally developed to measure the cumulative departure from equilibrium in surface water balance and its effect on soil moisture. It takes into account past and present precipitation, and potential evapotranspiration, via a two-layer bucket-type model for soil moisture calculations (section 12.7). The PDSI ranges from about -10 (dry) to $+10$ (wet), with values below -3 representing severe to extreme droughts. A PDSI time series reconstructed from tree-ring data for the Southwest United States is in Figure 12.1.

12.2 WHY DROUGHTS HAPPEN, CLIMATE TELECONNECTIONS

Droughts are typically associated with an anomalously high atmospheric pressure signal that persists during the precipitation season. Such a high pressure weather system has two effects that lead to drought. First, it diverts precipitation-bringing storms away from the area under the high pressure. This occurs because

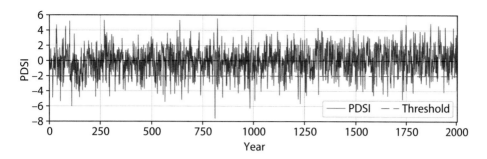

Figure 12.1: A long-term drought record.
A PDSI time series reconstructed for the Southwest United States from tree-ring data (section 12.4).

atmospheric winds travel around a high pressure (clockwise in the Northern Hemisphere) due to the Earth rotation and the resulting Coriolis force. The storms thus travel around the high pressure, leaving the area under its center dry. Second, high pressure centers tend to be associated with subsidence (sinking) motions. Subsidence, in turn, leads to the compression of the subsiding air, given that atmospheric pressure increases toward the surface, and as a result to its warming. The warming increases the saturation specific humidity following the Clausius-Clapeyron relation (Box 2.1), but its actual moisture content (specific humidity) does not change, leading to the decrease of its relative humidity. This drying reduces the capacity of moisture in the subsiding air to condense and precipitate and increases evaporation from the soil, further enhancing the drought. Finally, the subsidence-driven atmospheric drying also prevents condensation and cloud formation and increases the shortwave radiation reaching the surface, further increasing soil heating and drying.

There are several ways in which a high pressure over a given region interacts with subsidence in the context of droughts. First, as mentioned above, winds travel around a high pressure along equal pressure contours due to the Coriolis force (see also sections 4.2.1 and 8.1). However, if the high pressure extends all the way to the surface, the surface winds can develop a small component that flows from the center high pressure to the surrounding low pressure due to friction with the surface (this is the opposite of the situation involved in a flow around a low pressure, schematically shown in Figure 8.1 in the context of hurricanes). This flow out of the high pressure region is compensated by subsidence. A second subsidence mechanism involves an adiabatic flow along surfaces of constant dry static energy. Remember that dry static energy ($DSE = c_p T + gz$; see section 7.2) is conserved for adiabatic motion of air parcels not involving evaporation and condensation; therefore, air parcels in the mid-troposphere away from the atmospheric boundary layer (lower 1–2 km) move mostly along surfaces of constant dry static energy. The clockwise circulation driven in the Northern Hemisphere (anticlockwise in the Southern Hemisphere) by the high pressure center, combined with the mean atmospheric circulation in the region, can lead to a flow along a downward-sloping surface of a constant DSE and therefore to an adiabatic subsidence. As a result of this adiabatic subsidence, the height z is reduced for a given air parcel; therefore its temperature T must increase for its DSE to be conserved. This adiabatic compression and heating due to motion on sloping surfaces of constant dry static energy is an important mechanism for the intensification of drought conditions, desert conditions, and heat waves (chapter 13). Yet another final connection between subsidence and a high pressure center occurs as the air subsiding in the high pressure region experiences drying, which reduces condensation and therefore leads to clear-sky conditions. These clear-sky conditions allow air parcels to more effectively radiatively cool by emitting longwave radiation to outer space, as this radiation is not trapped and re-radiated downward by clouds. The radiatively cooling air parcels then become heavier and sink down, strengthening the subsidence. In the process of drought formation and maintenance, this radiative cooling (at a rate of about 1 K/day) is typically dominated by the heating due to the adiabatic compression discussed above.

High and low pressure centers due to weather events tend to travel and change on a timescale of a few days. A high pressure that persists at a given location for months and causes drought conditions must therefore be driven by some long-term forcing unrelated to weather, and the forcing is often due to a remote sea surface temperature anomaly (deviation from the long-term mean) that can be thousands of kilometers away. The SST anomaly forces large-scale atmospheric motions known as Rossby waves. The waves are seen as an alternating pattern of multiple high and low pressure centers in the mid-troposphere (say around 5 km height), which are separated by thousands of kilometers and extend all the way from the SST signal to the drought location. Such a persistent high pressure center forced remotely by Rossby waves can cause drought conditions. Because the SST anomalies that force the waves have timescales of months and longer, the remotely forced high pressure centers persist as well. The sea surface temperature anomalies may be due to climate variability modes such as El Niño/La Niña (Box 8.1), the Atlantic multidecadal oscillation (AMO; see chapter 6), the Indian Ocean dipole (Box 12.1), and more. The influence of remote SST anomalies on local conditions (temperature and/or precipitation) is referred to as *atmospheric teleconnections*; these are very common and influence many regions worldwide during different seasons. Heat waves (chapter 13) share many of the physical characteristics of droughts, except that they are weather events whose duration is much shorter and which do not require persistent remote forcing.

Climate Background Box 12.1
The Indian Ocean dipole

The Indian Ocean dipole (IOD) is a pattern of sea surface temperature in the equatorial Indian Ocean that alternates irregularly between the positive and negative phases shown in panels a and b of the accompanying figure. The time series in panel c shows the IOD index calculated as the averaged sea surface temperature anomaly over the west minus east Indian Ocean SST (green boxes, 50°E–70°E, 10°S–10°N minus 90°E–110°E, 10°S–0°). Unlike the El Niño/La Niña variability in the Pacific (Box 8.1), the IOD may not reflect an independent variability mode in the Indian Ocean and is likely at least partially driven by the El Niño–La Niña/ENSO cycle, although details of this interaction between ENSO and the IOD are still being studied. The Indian Ocean SST pattern still has a role in forcing remote teleconnections affecting precipitation, drought conditions, forest fires, and more over large areas. The IOD and ENSO have complex and irregular spatiotemporal structures, are characterized by a range of temporal and spatial scales, and are affected by numerous physical processes. The difficulties in diagnosing the relation between the IOD and ENSO demonstrate how the relatively short well-observed climate record of the past decades is often insufficient to answer some key questions about inter-annual and decadal climate variability whose timescale requires a longer observational record to separate a clean signal from noisy observations.

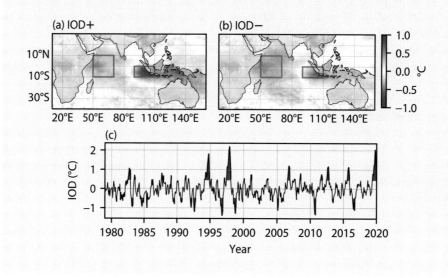

An example of how the above factors combine to create January drought conditions in California in a long climate model run is in Figure 12.2. The left panels show averaged January conditions (climatology) for precipitation, atmospheric pressure at sea level, and sea surface temperature. The right panels show the deviations (anomalies) from these climatologies, averaged over Januaries that are exceptionally poor in California precipitation (note blue patch centered near California in Figure 12.2b), representing severe meteorological drought conditions there. The years corresponding to these drought conditions are also characterized by a strong high sea level pressure signal over California (Figure 12.2d) and cool East Equatorial Pacific sea surface temperature, corresponding to a La Niña SST pattern (Figure 12.2f; see also Box 8.1) that is remotely forcing the drought conditions. Careful analysis of observed droughts over California suggests that they are driven by a diverse set of factors in addition to La Niña SST.

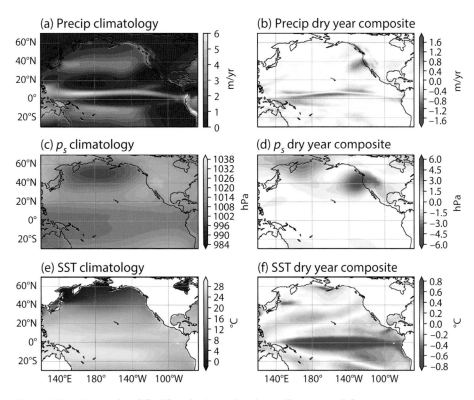

Figure 12.2: Analysis of California droughts in a climate model.
The coupled ocean-atmosphere model is run at a preindustrial CO_2 concentration: climatological January averages (*left*) and the deviations from these climatologies averaged over dry California Januaries (*right*) for precipitation (*top*), sea level pressure (*middle*), and SST (*bottom*).

12.3 DETECTION OF CLIMATE CHANGE

In order to explore how unusual a given decade that experienced a few drought years is, one can use *non-parametric* statistical analysis. This is a simple yet effective tool that is used for the analysis of extreme events in general and is also very useful for the detection of climate change. Consider a long time series of the annually averaged PDSI for the Southwest United States (Figure 12.1). There are three drought years (defined for the purpose of the present analysis as PDSI < -2) in the last three decades of the record, which only extends to year 2000, and we wish to find out how unusual such a period is. Could it be just a coincidence rather than a signal of anthropogenic climate change? To find out, shuffle the points in the annual time series randomly, and calculate the

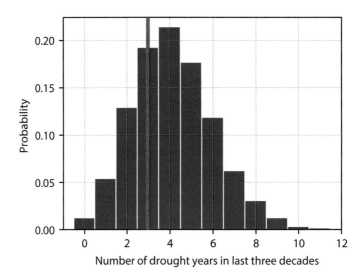

Figure 12.3: Detection of climate change.
A probability distribution function of the number of drought years within the last three decades, obtained by repeatedly randomly shuffling the PDSI record in Figure 12.1.

number of drought years in the last decades of the shuffled time series. Repeat this many, say 10,000, times (realizations), and calculate the probability distribution function (PDF) of the number of drought years. That is, calculate the number of occurrences of a given number of drought years in the last decades as a function of the number of drought years in the last decades, normalized by the total number of events; see Figure 12.3. As shuffling removes any trends from the time series, the number of occurrences of drought events in the last decades of the shuffled time series represents the possibility that these droughts are coincidental rather than the result of a recent trend due to anthropogenic global warming.

We can now compare the observed number of events in the last decades of the non-shuffled data (gray bar in Figure 12.3) to the PDF to find out if the observed number of events in the last decades rarely occurs. Based on the bars in the figure, we calculate the probability of encountering eight or more drought years in three decades, for example, by summing the bars representing eight or more drought years and find that the probability is less than 5%. This indicates that such an occurrence may be considered unusual, because the assumption that the number of events during the past few decades is part of the normal variability was rejected with a 95% confidence level. In other words, had we encountered eight events or more in the last three decades of the record, we would have detected the signature of climate change with a statistical significance of 95%. Based on this analysis, the observed number of drought events indicated by the gray bar in Figure 12.1 does not seem to be exceptional as the probability of encountering three or more events is not small. An analysis based on extending the data of Figure 12.1 to recent decades suggests that California droughts in the 21st century may have been exceptionally severe.

12.4 OBSERVATIONS, PALEO PROXY DATA

As shown above, having a long record of drought conditions, say based on the PDSI, is a critical tool for evaluating how unusual recent decades are and for deciding if the occurrence of drought years may be attributed to climate change or is a result of natural variability. Tree-ring width data are commonly used to reconstruct droughts over the past couple of thousands of years. Tree rings record the seasonal growth season and tend to be wider for rainier seasons. The challenges are that tree-ring widths need to be empirically related to precipitation rates, the relation is different for different tree species, and the signal is very noisy, as seen in Figure 12.4, making it necessary to average over a very large number of tree-ring records in order to extract a climate signal from the noise.

Such long records allow us to identify major past drought events such as the medieval megadroughts in North America. These multiple extreme drought

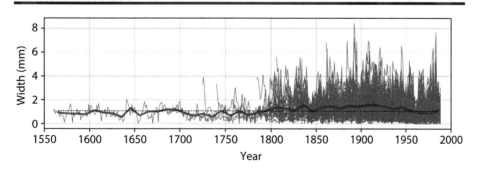

Figure 12.4: Tree-ring data.
Multiple tree-ring width data for bigcone Douglas-fir trees from the San Bernardino Mountains (gray lines), plotted with their decadal bin-average (red line) and its mean (dashed line).

events lasted decades each, affecting western North America between about AD900 and 1400, and the record suggests that the medieval climate of the West was significantly more arid than the following centuries and than the current climate. Tree-ring records in this case are supplemented by other indications of dry climate, such as dead tree stumps that grew in dry areas that are now flooded by lakes and the number of bison bones in archaeological sites. These persistent drought events seemed to have been part of a global pattern that included wet conditions in northern South America, dry climate in mid-latitude South America and in eastern Africa, and more. Together, these patterns suggest a possible driving by persistent La Niña conditions in the tropical Pacific. The reconstruction of these dramatic past events suggests that the climate system can operate in ways that are different from what we are familiar with and understand.

12.5 EXAMPLE PROJECTIONS: SOUTHWEST UNITED STATES AND THE SAHEL

We have seen that droughts often occur due to persistent deviations of the sea surface temperature from its long-term mean, such as caused by the natural variability due to El Niño/La Niña (Box 8.1) or due to the Indian Ocean dipole (Box 12.1). Because our ability to predict the response of these natural SST variability modes to global warming is currently limited, it is also difficult to predict the future statistics of droughts in particular regions. This difficulty is supplemented by the uncertainty involved in predicting precipitation more generally, as it often depends on atmospheric convection, which is challenging to accurately represent in climate models, as we have seen in section 7.2, and as manifested by the disagreement on precipitation projections among different climate models. We now demonstrate this difficulty by considering two regions, the Sahel and the Southwest United States.

Droughts in the Sahel, south of the Sahara, are famously disastrous to its tens of millions inhabitants. Persistent droughts occurred there toward the end of the 20th century, although it is not clear that this trend still persists. Future RCP8.5 projections for the Sahel are divided into some that show no change in a significantly warmer climate and others that show the Sahel getting *wetter*. This is demonstrated in Figures 12.5a–c, each of which show the projections of

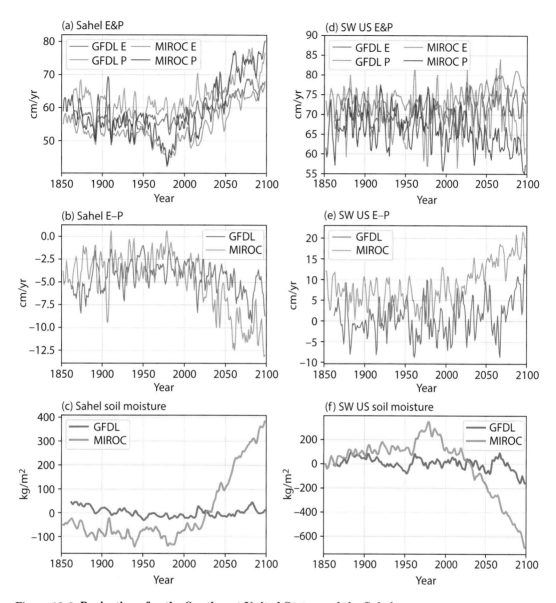

Figure 12.5: Projections for the Southwest United States and the Sahel.
Historical and projected (RCP8.5) drought-related model variables for 1850–2100 for two climate models, GFDL and MIROC, for the Sahel (a–c) and for the Southwest United States (d–f).

two climate models. The evaporation (E) and precipitation (P) over the Sahel in panel a seem noisy, and differences between the two models are not obviously apparent. The evaporation minus precipitation in panel b shows some differences, where the model shown by the orange curve calculates a larger precipitation excess toward the end of the 21st century. Finally, the soil moisture anomaly projections, which accumulate differences in evaporation, precipitation, and runoff over time, show that one model predicts a vast increase in soil moisture while the other predicts that it remains similar to its present-day values. Section 12.7 discusses how soil moisture depends on evaporation and precipitation, and in particular how small differences in precipitation rate can lead to an amplified response of soil moisture.

The Southwest United States experienced some remarkable droughts in the first decades of the 21st century. Future projections of multiple climate models seem to more consistently (than for the Sahel) show a persistent drying of this region in a future warmer climate, although the model disagreement and thus uncertainty in these projections are still large, as clearly demonstrated by the two models in Figures 12.5d–f. Interestingly, paleoclimate evidence suggests that the Southwest United States was wet during the most recent warm climate period of the Pliocene (Box 4.1).

12.6 UNDERSTANDING PRECIPITATION TRENDS

An important goal of climate change science is to predict and understand changes to the global and regional distributions of precipitation. Equally important is the prediction of precipitation minus evaporation, which determines drought conditions and soil moisture, which are of utmost important to critical human activities. Note first that remote forcing by appropriate SST patterns via a Rossby wave train can lead to excess precipitation, just as (other) SST patterns were shown above to lead to droughts. If the Rossby waves lead to a persistent low pressure over a given area, this will lead to rising motions and, as discussed in chapter 7, therefore to adiabatic cooling, condensation, and a regional signal of excess precipitation, exactly the opposite of the mechanism for droughts discussed above. As a result, if SST patterns change in a future climate, some regions might experience persistent floods, just as others might see persistent droughts. This section complements the above discussion of regional changes to droughts or excess precipitation, with a study of global-scale trends in precipitation.

We take here two complementary approaches to understanding global-scale projections of precipitation patterns. First, in section 12.6.1 we discuss the projected expansion of the atmospheric Hadley cell, which may affect the future location of desert bands; second, in section 12.6.2, we examine the projection that precipitation changes will lead to a "wet getting wetter, dry getting drier" scenario, including its limitations. Finally, while the mean precipitation is highly relevant to many socioeconomic activities, the distribution of precipitation in time can also play an important role. That is, whether precipitation falls in a few intense events or slowly accumulates over time makes a difference, as discussed in the introduction to this chapter. In section 12.6.3 we show and explain future projections that more of the total precipitation is expected to occur in extreme events in a warmer climate.

12.6.1 Hadley cell expansion and weakening

The strong insolation at the equator leads to the warming and therefore rising of air there, which then travels poleward in both hemispheres and sinks around 30° north and south. This circulation is known as the Hadley cell, and it is a major player in setting global-scale precipitation patterns. The rising motions near the equator lead to adiabatic cooling, condensation, and rain, while the desert belts in both hemispheres (Sahara, Kalahari, and others) are often attributed to the downward, subsidence, motions of the Hadley cell, which lead to warming and drying. Any changes to the amplitude or spatial structure of this atmospheric circulation in response to greenhouse gas increase can have significant implications to global precipitation patterns.

Figure 12.6 shows contours of the streamfunction of the annually averaged, zonally averaged meridional atmospheric circulation. The zonally averaged flow occurs along these contour lines (as was the case for the Atlantic meridional overturning circulation; see chapter 6). Blue and red colors show the streamfunction evaluated for 1920–1940, and gray contours show it for the RCP8.5 scenario, averaged over 2080–2100. Blue colors (and dashed gray contours) indicate *anticlockwise* flow, and red shades (solid gray lines) indicate clockwise motions, so that the figure represents air rising over the equator and sinking at

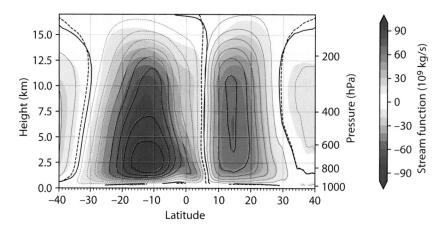

Figure 12.6: Projected Hadley cell expansion under global warming.
The color shading shows the zonally averaged atmospheric meridional overturning circulation evaluated by a climate model for 1920–1940. The contour shading levels indicate the transport in Sverdrups (Sv, defined here as 10^9 kg/s). The gray lines show the same quantity for an RCP8.5 scenario, averaged over 2080–2100. The solid black contour denotes the zero of the 1920–1940 streamfunction, and the dashed black contour is that for 2080–2100.

around $30°$ north and south. The mass transport carried by the atmospheric circulation between two streamfunction contours is equal to the difference in the contour values. Therefore, the total transport carried in the Southern Hemisphere at about $12°$S between a height of 2.5 km and 15 km is about 100×10^9 kg/s, with similar transport carried by the Northern Hemisphere Hadley cell.

Two main results are apparent when comparing the color shading and gray contours. First, the circulation is projected to weaken. The maximum streamfunction values in the Northern and Southern Hemispheres, during 1920–1940, are 70 and 92 Sv, correspondingly. For 2080–2100, these maximum values are weakened to 62 and 89 Sv. Second, the poleward edge of the downward circulation shifts poleward. This is indicated in Figure 12.6 by the solid and dashed black contour lines, which show the zero streamfunction value contours and thus mark the edges of the Hadley cells, for 1920–1940 and 2080–2100, correspondingly. Note that the dashed black line in the Southern Hemisphere, at around 5 km altitude, is shifted by over a degree latitude (one degree latitude corresponds to about 110 km) poleward relative to the solid black line. A similar, though weaker, poleward shift is also seen in the Northern Hemisphere.

We next attempt to rationalize the Hadley cell weakening using a moisture budget for the lower 1–2 km of the atmosphere (the *boundary layer*). For this purpose, remember that the tropical near-surface atmospheric humidity increases by about 6% per degree of surface warming, assuming the relative humidity does not change (based on the Clausius-Clapeyron relation; see Box 2.1), or by about 20% for a three-degree warming representing a mid-range RCP8.5 projection for the tropics. The moister boundary layer air implies that an upward air mass transport in the tropics in a warmer climate, that is equal to its preindustrial value, would carry with it about 20% more moisture out of the boundary layer. At the same time, the surface evaporation rate is known to be less sensitive to warming, so that the increased upward transport of more moisture is not balanced by a corresponding increase in the input of moisture from surface evaporation, and this suggests that the boundary layer would dry out. That is, an increased moisture content of the boundary layer is not consistent with the Hadley circulation remaining at its preindustrial strength. To maintain a steady state of the boundary layer moisture content, the Hadley cell transport must weaken, so that the net moisture transport does not increase. This qualitative argument is consistent with the weakening seen in Figure 12.6, although see the caveat below regarding the fuller explanation of this weakening.

The poleward expansion of the poleward edge of the Hadley cells is of special interest due to its possible effects on the extent of the desert belts, and it can be explained via the stability of the tropospheric jets at the edge of the Hadley cells. An air parcel rising in the present climate at the equator and traveling poleward at an altitude of a few kilometers is shifted by the Coriolis force to the right in the Northern Hemisphere and to the left in the Southern Hemisphere, creating the subtropical upper-level westerly (eastward) tropospheric jets. These jets become stronger the further the air moves poleward, eventually become unstable, and break into weather-scale motions, not allowing the jets to further strengthen, and setting the poleward edge of the Hadley cell. We saw previously that in a warmer climate, the tropical lapse rate weakens (Figure 3.6) due to the way moist convection operates (section 7.2). It turns out that this change to the lapse rate allows the Hadley cell to further expand poleward before becoming unstable, explaining the poleward expansion seen in Figure 12.6. It should be stated that the above moisture budget argument and the mechanisms for both the weakening and expansion of the Hadley cell are only partial as well

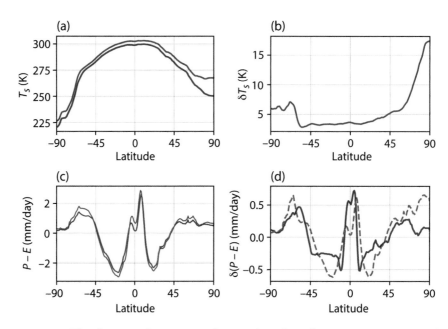

Figure 12.7: The "wet getting wetter, dry getting dryer" response to a warming scenario.
(a) Zonally averaged surface temperature as a function of latitude during 1920–1940 (blue) and 2080–2100 (red) in the RCP8.5 scenario. (b) The net surface warming as a function of latitude. (c) Zonally averaged $P - E$ for the same year ranges. (d) The change in $P - E$ (solid) vs its predicted structure based on equation 12.2 (dashed).

---

as an over-simplification of the fuller explanations explored in the recent scientific literature. Additional factors in play in the Hadley cell weakening and expansion include changes to the atmospheric lapse rate, tropopause height, and others.

12.6.2 "Wet getting wetter, dry getting drier" projections

Figure 12.7c shows the time-mean zonally averaged precipitation minus evaporation $(P - E)$ as a function of latitude estimated for 1920–1940 (blue). It also shows (red) the zonally averaged precipitation minus evaporation, projected for the end of the 21st century, in response to the warming shown in panels a and b under the RCP8.5 scenario. The general pattern is that in latitudes where $P - E$ is positive in the preindustrial climate, it becomes even more positive in the warmer climate ("wet getting wetter"), while where it is negative, it becomes even more so ("dry getting drier").

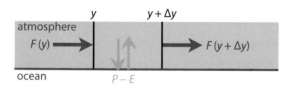

A schematic of moisture budget of a slice of the atmosphere bounded by two latitudes y and $y + \Delta y$, used to derive the relation (12.1) between precipitation minus evaporation (green arrows) and the convergence of the poleward moisture flux $F(y)$ (red).

To understand this interesting pattern of projected precipitation response to warming, we examine the poleward transport of moisture by the atmosphere, and its dependence on warming, and for now ignore zonal variations in the related atmospheric variables. First, the precipitation minus evaporation as a function of the northward distance y (say measured in m from the equator) is related to the atmospheric poleward flux of moisture per unit distance in the east-west direction, $F(y)$. This moisture flux is calculated as the poleward air velocity times the atmospheric moisture times air density, integrated over height, and is given in units of kilograms of moisture per second per meter. The relationship between this moisture flux and precipitation minus evaporation ($P - E$, in units of kg moisture per second per m^2; when plotted as in Figure 12.7, one often divides this flux by water density to convert the units to those of velocity, say mm/day) may be written as

$$(P - E) = -\frac{dF}{dy}. \tag{12.1}$$

This may be understood via Figure 12.8, describing the moisture budget of a slice of atmosphere extending from y to $y + \Delta y$. The moisture flux out of the slice due to precipitation minus evaporation, per unit distance in the east-west direction, is $(P - E)\Delta y$. For the moisture content of the slice to be in a steady state, the moisture flux carried by poleward winds into the slice at y, minus that out of the slice at $y + \Delta y$, minus the flux out of the slice due to precipitation minus evaporation from the underlying ocean or land surface, should vanish. This implies $F(y) - F(y + \Delta y) - (P - E)\Delta y = 0$, and dividing by Δy and using the definition of a derivative, we find equation (12.1).

The poleward moisture flux F per unit distance in the east-west direction is largely due to a poleward air volume flux per unit distance in the east-west direction (m² per second) in the lower atmosphere, where most of the moisture resides, times the specific humidity of the air (kg moisture per kg moist air) times the air density (kg air per m³). As the atmosphere warms, its moisture content increases following the Clausius-Clapeyron relation (Box 2.1). We assume that the changes to the air density, and to the atmospheric circulation and therefore to the poleward lower atmospheric air mass flux, are small. As a result, the change to the meridional moisture flux due to warming, δF, is assumed equal to the change in moisture content times the unchanged poleward air mass flux (volume flux times air density). The relative change to the poleward moisture flux, $(\delta F)/F$, is therefore equal to the relative change in the specific humidity, $(\delta F)/F = (\delta q)/q$. Assuming the relative humidity remains approximately constant as the atmosphere warms, we can write $(\delta F)/F = (\delta q^*)/q^* = \alpha_{cc}\delta T$, where q^* is the saturation specific humidity and α_{cc} is the expected relative change to the saturation moisture, per degree Kelvin warming, due to the Clausius-Clapeyron relation as explained in Box 2.1. This implies $\delta F = \alpha_{cc}\delta T \times F$. Equation (12.1) implies that the change in $P - E$ due to warming is proportional to the divergence of the change in moisture flux, $\delta(P - E) = -d(\delta F)/dy$. Substituting the expression we found for the change in the flux, we find

$$\delta(P - E) = -\frac{d}{dy}(\delta F) = -\frac{d}{dy}(\alpha_{cc}\delta T \times F) \approx -\alpha_{cc}\delta T \times \frac{d}{dy}(F)$$

$$= \alpha_{cc}\delta T \times (P - E),$$

where we assume in the third equality that the warming δT is not a strong function of latitude (an acceptable assumption away from the polar areas), so we can take it out of the derivative. The final result is

$$\delta(P - E) = \alpha_{cc}\delta T \times (P - E), \tag{12.2}$$

which implies that the change in the precipitation minus evaporation, $\delta(P - E)$, due to global warming is proportional to its preindustrial pattern. This means that when $P - E$ is positive, it is expected to be even larger, and when it is negative (tendency to dry conditions), it is expected to become more negative. As evaporation depends on temperature less strongly than precipitation, one might use these predictions to also deduce changes to precipitation in addition to $P - E$.

The above arguments therefore explain the "wet getting wetter, dry getting drier" pattern seen in climate models (Figure 12.7). The proportionality constant in equation 12.2 is $\alpha_{cc}\delta T$, which would correspond to a change of some 6% per degree Kelvin of warming.

It is important to keep in mind what assumptions were made in order to derive equation (12.2): (1) the relative humidity will not change significantly in a warming scenario, (2) the main change to the meridional atmospheric moisture flux is due to increased moisture rather than to wind changes, and (3) the warming is relatively uniform as a function of latitude away from the poles.

It is remarkable that we can understand changes to the distribution of precipitation minus evaporation using such simple arguments. Yet it turns out that these arguments work well mostly over the ocean but less well over land, where evaporation is not unlimited as it is over the ocean, and where the teleconnections discussed above seem to play a more dominant role. As a result, predictions of precipitation and of $P - E$ over land, where the potential socioeconomic impact is especially great, largely do not follow the pattern of "wet getting wetter, dry getting drier" and are much less certain, as reflected in significant differences among the projections of different models.

12.6.3 Precipitation extremes in a warmer climate

Consider a projection of extreme precipitation events in the RCP8.5 warming scenario in Figure 12.9, comparing the periods of 1920–1940 and 2080–2100. Examine all days during each of these two periods, and sort the daily precipitation rates for a given latitude by their magnitude. Then find the extreme precipitation rate that is larger than the rate in 99.9% of all days. Repeat for all latitudes and for both 1920–1940 and 2080–2100; the results are in Figure 12.9a. The blue and red curves, showing the extreme precipitation rates for 1920–1940 and 2080–2100, correspondingly, suggest that a significant increase in the magnitude of the daily precipitation corresponding to this percentile is expected by the end of this century. That is, the precipitation rates that occur in extreme precipitation events are more intense in a warmer climate.

The same conclusion is seen in the more detailed analysis of precipitation at latitude $40°N$ in Figure 12.9c. The two curves show the cumulative

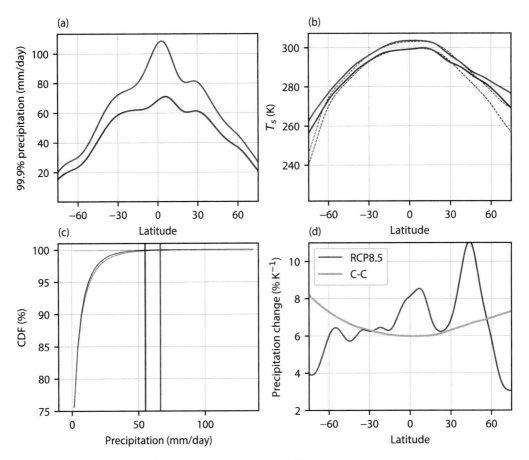

Figure 12.9: Extreme precipitation events in the RCP8.5 scenario.
(a) The intensity of the 99.9th percentile of daily precipitation rates as a function of latitude, as simulated by a climate model for 1920–1940 (blue) and for 2080–2100 (red). (b) The dashed lines show the zonally averaged surface temperature for the same two periods, while the solid lines show the same temperature during extreme precipitation events. (c) The cumulative distribution function for daily precipitation at 40°N for the two periods, with vertical bars showing the precipitation rates corresponding to the 99.9th percentile. (d) The fractional increase in the magnitude of the 99.9th percentile of daily precipitation rate, as a function of latitude, per degree K warming at the same latitude, in % K^{-1} (red). The gray line shows the Clausius-Clapeyron projection.

probability distribution function, which is calculated as follows. Consider the probability distribution function for precipitation events, $f(P)$, where P is the daily precipitation rate (mm/day), such that $f(P)dP$ is the probability of the daily precipitation rate being between the values $(P, P + dP)$. The cumulative probability distribution function is then defined as the integral over the probability distribution function, $CDF(P) = \int_{-\infty}^{P} f(P')dP'$. Its value at a given precipitation rate is the probability of the precipitation rate being less than that value. Figure 12.9c shows the CDF curves for precipitation at 40°N for the two periods, and while the difference seems small overall, the vertical lines show the precipitation rates corresponding to a probability of 99.9% for the two periods, indicating a significantly stronger extreme precipitation rate in a warmer climate (red) versus preindustrial (blue).

A useful measure of the change to extreme precipitation events in response to warming is the percent change in the precipitation rate corresponding to the 99.9th percentile at a given latitude, per degree Kelvin of warming at that same latitude (red solid line in Figure 12.9d). The change is a strong function of latitude, with extratropical values being generally around 3%–6% per Kelvin. While the particular model used for this analysis also shows stronger intensification in the tropics, this turns out to be a less reliable model prediction due to large differences in the tropics between different models, as explained shortly. In any case, the projected few percent warming per degree in the extratropics can accumulate to a significant change for an RCP8.5-like scenario of a few degrees warming.

To understand these projections for precipitation extremes, consider the process by which precipitation typically forms. As we saw in section 7.2, a rising air parcel encounters lower pressure levels and therefore expands and cools adiabatically, leading to condensation and to the precipitation of some of the condensed water. A first attempt to estimate the increase in extreme precipitation may therefore be based on the fact that saturation humidity increases exponentially with temperature based on the Clausius-Clapeyron relation (Box 2.1). Assuming the relative humidity does not change significantly as the atmosphere warms, one expects higher water content of rising air parcels in a warmer atmosphere, and therefore higher precipitation rates. Figure 12.9b shows the surface temperatures averaged over 1920–1940 (dashed blue line), for 2080–2100 (dashed red line), and during extreme precipitation events (solid lines), showing that extreme precipitation events in the extratropics happen during warmer days than the mean for each period (compare dashed to solid lines). Given the saturation specific humidity as a function of temperature T and pressure p, $q^*(T, p)$, we estimate the percent change with temperature as $100 \cdot (dq^*(T, p)/dT)/q^*(T, p)$. This estimate, referred to as the Clausius-Clapeyron scaling, is shown for a typical surface pressure as the gray line in Figure 12.9d. The resulting curve is on the same order of magnitude as the actual percent change to extreme precipitation rates calculated from the climate model (red curve), confirming the idea that enhanced extreme precipitation is a result of increased moisture content of air parcels due to warming. However, the Clausius-Clapeyron scaling is clearly an overestimate outside the tropics, and we attempt to understand the reason for that next.

A further attempt to refine the estimate of increased precipitation rate in extreme events, P_e, may be derived as follows. As the parcel rises, it reaches saturation at the lift condensation level (LCL). Above that point, a further decrease in the saturation specific humidity leads to condensation, and the condensation rate is equal to the rate of decrease in the saturation specific humidity of the parcel, $-dq^*/dt$. The specific humidity decreases as the parcel rises, while the condensation rate is positive—hence the minus sign. The actual contribution to surface precipitation rate is $-\epsilon_p dq^*/dt$, where ϵ_p is the *precipitation efficiency*, the fraction of the condensed water that precipitates to the ground, as opposed to being re-evaporated or horizontally transported away from the condensation site.

The rate at which a parcel's saturation specific humidity decreases depends on how rapidly the parcel rises and experiences lower pressure levels. Write this rate of decrease in the saturation specific humidity as $dq^*/dt = (dq^*/dp) \times (dp/dt)$. The rate of change of pressure with time is referred to as the pressure vertical velocity; it is denoted $\omega = dp/dt$ and is negative for a rising air parcel as the pressure decreases upward. However, we remember from section 7.2 that as the moisture in the parcel condenses, latent heat is released and thus lowers the parcel's cooling rate, which also lowers the rate at which the saturation specific humidity changes. In the process, moist static energy is conserved, so we need to take the derivative of the saturation specific humidity with respect to pressure at a constant MSE. The total extreme precipitation rate at the surface is then the vertical integral over the precipitation efficiency times the condensation rate, written as

$$P_e = - \left\{ \epsilon_p \omega \left. \frac{dq^*(T,p)}{dp} \right|_{MSE} \right\}. \qquad (12.3)$$

In this equation, the curly brackets denote the vertical integral over the path of the rising parcel. The condensation rate of the rising parcel is given by the pressure vertical velocity times the rate of change of the saturation specific humidity with pressure, with the MSE kept constant. Finally, the precipitation efficiency factor accounts for the fact that not all the condensed water precipitates to the ground.

We can now understand the possible factors that can lead to a change in the extreme precipitation rate in a warmer climate. First, the precipitation efficiency may change. This factor is affected by the cloud microphysics (section 7.3) and therefore is referred to as the *microphysical* component; it is affected by processes that occur on a very small scale, and its dependence on atmospheric warming is highly uncertain. The second factor is the vertical velocity during extreme precipitation events, or the *dynamical* component. As noted in chapter 7, this vertical velocity may be due to wind blowing over a mountain, to vertical velocities that might develop in mid-latitude weather systems, or to updrafts in convection events. In the tropics, the vertical velocity is dominated by convective updrafts that occur over a small horizontal length scale of a few hundred meters to a few kilometers. These motions cannot be explicitly resolved and simulated by climate models whose horizontal grid spacing is much larger, so they are instead calculated by sub-grid-scale convection parameterizations (as discussed in chapter 7). Because different climate models use different representations (parameterizations) of convection and convective updrafts, the responses to climate warming of the vertical velocity in tropical convection events vary significantly among models. This explains the above-mentioned observation that the projection of tropical extreme precipitation in a warmer climate varies significantly among models and is therefore very uncertain. Even away from the tropics, the dynamical component is due to a very diverse set of processes, and it is difficult to robustly predict its response to warming.

The third factor in equation 12.3 is the *thermodynamic* component of the estimate for the extreme precipitation rate, which may be written somewhat less cryptically as

$$\frac{dq^*(T,p)}{dp}\bigg|_{\text{MSE}} = \frac{\partial q^*(T,p)}{\partial p} + \frac{\partial q^*(T,p)}{\partial T}\frac{\partial T}{\partial p}\bigg|_{\text{MSE}}.$$

The final factor here is related to the moist adiabatic lapse rate $\partial T/\partial z$ taken at a constant MSE, which was calculated in section 7.2,

$$\frac{\partial T}{\partial p}\bigg|_{\text{MSE}} = \frac{\partial T}{\partial z}\bigg|_{\text{MSE}}\frac{\partial z}{\partial p},$$

where $\partial z/\partial p$ can be calculated from the exponential dependence of the pressure on height (section 7.2).

Figure 12.10 compares the above two attempts to estimate changes to extreme precipitation rates as a function of warming: the saturation specific humidity (blue) and the thermodynamic component of the above estimate, $dq^*(T,p)/dp$ at a constant MSE (orange), both drawn as a function of temperature. The thermodynamic component clearly shows a smaller sensitivity to temperature than the straightforward scaling based on the saturation specific humidity. This explains why model projections for extreme precipitation events away from the tropics seem to be somewhat less dramatic than might be expected from pure Clausius-Clapeyron scaling (red vs gray curves in Figure 12.9d). Recent studies making a more careful use of expression 12.3, by carrying out the integral implied there, find a good fit to the projected increase in extreme precipitation events by full complexity climate models. Such a good fit strengthens the case for the physics behind this expression and thus for our understanding of, and confidence in, the expected increase in extreme precipitation rates.

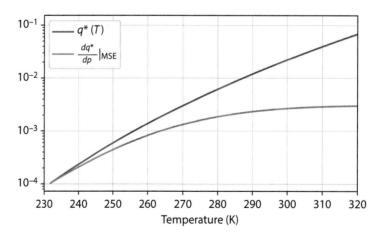

Figure 12.10: Two estimates for the dependence of extreme precipitation rates on temperature.
One is based on the saturation specific humidity (Clausius-Clapeyron scaling, blue) and the other on the thermodynamic component of the estimate from equation (12.3). The orange curve is scaled by a constant factor to have the same value as the blue one at the lowest temperature plotted.

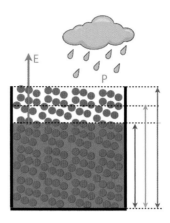

Figure 12.11: A bucket model for soil moisture. Schematic of the model, showing the saturated soil depth range (blue) under the unsaturated range.

When available precipitation observations are analyzed in order to identify an observed change to extreme precipitation events in different regions of the globe, one sees both increases and decreases in the magnitude of such events over the past decades, although there are more regions showing increase than decrease. As seen in the above arguments, the expected change is on the order of a few percents per Kelvin in the magnitude of the 99.9th percentile precipitation rate. This is a relatively small signal for the warming that has occurred so far, and such events are rare by definition, making detection and attribution of an observed signal difficult given the short, noisy, and highly variable—in both space and time—character of precipitation records. But the physical arguments above seem to suggest that an increase in extreme precipitation is a robust projection in many regions, at least outside the tropics.

12.7 A BUCKET MODEL FOR SOIL MOISTURE

In order to understand and predict soil moisture deficits that lead to hydrological droughts, we should consider how precipitation and evaporation affect soil moisture. The simplest approach to predicting soil moisture is based on a *bucket model* (schematic Figure 12.11), where the soil is treated as a bucket accumulating precipitating water and losing it via evaporation. While state-of-the-art land models within modern global climate models use a more sophisticated formulation, the principles demonstrated here still apply.

First, the evaporation from a very wet soil (potential evaporation, m/s), E_0, is given by equation (12.4), where ρ_{air} is the air density (kg/m^3) at the surface, ρ_{water} is water density, C_k is a nondimensional empirical *bulk* *coefficient*, $q^*(T)$ the saturation specific humidity (kg water vapor mass per kg of moist air) at surface atmospheric pressure and a surface temperature T, q the atmospheric surface specific humidity, and $|\mathbf{V}|$ the surface wind velocity vector magnitude (m/s).

Soil moisture, the amount of water stored in the soil, can be quantified by the soil water content in the upper 1 m of soil, W, which is the height of a water column (in meters), had it been extracted from the soil. If soil water content is low, moisture is stored below the surface in the saturated part of the soil, up to the groundwater table (red arrow in Figure 12.11), lowering the rate of evaporation to the atmosphere due to the *resistance* of the unsaturated soil above. The resistance is a result of the water vapor needing to diffuse from the groundwater table (uppermost level of saturated soil) to the soil surface. A lower evaporation rate due to soil resistance, calculated in equation (12.5), is expressed by the factor W/W_k applied to the evaporation rate, which becomes smaller for small moisture levels W. The constant W_k (schematically indicated by the green arrow in Figure 12.11) is the soil moisture level below which evaporation rate begins to be suppressed by soil resistance. The maximum possible amount of water content in the upper 1 m of soil (brown arrow, denoted W_{FC} and measured as the height of the water column extracted from 1 m of a water-saturated soil) is determined by the soil porosity and depends on the soil type and grain size. For the calculations below, we set the maximum water content to an example value of $W_{FC} = 15$ cm, thus assuming a 15% porosity, and similarly estimate the threshold soil moisture beyond which resistance to evaporation becomes activated at $W_k = 0.75 W_{FC}$.

Finally, denoting the precipitation rate by P, we can write the budget equation for soil moisture (eqn 12.6). The complete bucket model equations are as follows:

$$E_0 = \rho_{air} C_k |\mathbf{V}| (q^*(T) - q)/\rho_{water} \tag{12.4}$$

$$E = \begin{cases} E_0 & \text{if } W \geq W_k \\ E_0 \dfrac{W}{W_k} & \text{if } W < W_k \end{cases} \tag{12.5}$$

$$\frac{dW}{dt} = \begin{cases} 0 & \text{if } W \geq W_{FC} \text{ and } P - E > 0 \\ P - E & \text{otherwise} \end{cases} \tag{12.6}$$

Note that in the first line of equation (12.6), if the water content is equal to its maximum value of W_{FC} and precipitation minus evaporation is positive, therefore tending to further increase the water content beyond its possible maximum, it is assumed that the excess water is lost to runoff and therefore dW/dt is set to zero, and the soil water content does not further increase. This limits soil moisture to always be less than or equal to W_{FC}.

Example

Consider first an estimate of potential evaporation for specific meteorological conditions representing hot and dry conditions in an arid or semi-arid region. Standard parameter values could be taken as $\rho_{air} = 1.225$ kg/m^3, $\rho_{water} = 1000$ kg/m^3, and $C_k = 1.0 \times 10^{-3}$. Using equation (12.4), assume 60% atmospheric relative humidity so that $q^*(T) - q = (1 - q/q^*(T))q^*(T) = (1 - RH)q^*(T) = 0.4q^*(T)$, and let the temperature be 32 °C and the wind velocity 10 m/s. To convert the evaporation to units of meter water per year, divide by the density of water and multiply by the number of seconds in a year. Using a saturation specific humidity of 30 gr/kg, we find

$$E_0 = \frac{\rho_{air}}{\rho_w} C_k |\mathbf{V}|(1 - RH)q^*(T)$$

$$= \frac{1.225 \, (\text{kg/m}^3)}{1000 \, (\text{kg/m}^3)} \times 10^{-3} \times 10 \, (\text{m/s}) \times 0.40 \times 0.03 \, (\text{kg/kg})$$

$$\times (3600 \times 24 \times 365) \, (\text{s/year}) = 4.6 \, \text{m/yr}.$$

Such a large potential evaporation rate in a semi-arid region would not be compensated by precipitation and leads to the dry conditions in such regions.

The bucket model may now be used to deduce how evaporation and soil moisture adjust to equilibrium in two scenarios in Figure 12.12. In both cases we assume that the initial soil moisture is 0.14 m. Then we first (blue curves) set the precipitation rate to 2.2 m/yr and see that the soil water content gradually decreases and equilibrates at $W = 10.7$ cm (panel a). Panel b shows how the precipitation minus evaporation approaches zero during the equilibration, as expected. Because precipitation is prescribed, this evolution represents an adjustment of the evaporation rate. Note how initially the evaporation does not change (blue line in panel b is initially horizontal). When the decreasing water

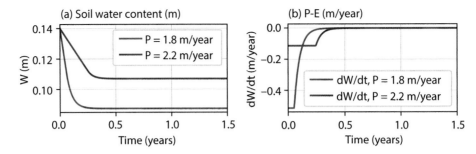

Figure 12.12: Results of the bucket model for two prescribed precipitation rates, as a function of time.
(a) Soil water content. (b) RHS of the soil moisture equation (12.6).

content reaches the threshold of $W_k = 0.75 W_{FC} = 0.1125$ m and the evaporation starts decreasing due to the soil resistance factor W/W_k, the precipitation minus evaporation (blue line) gradually increases to zero, where an equilibrium soil water content is reached.

Next, reduce the precipitation rate to a lower value of 1.8 m/yr, and note that the soil moisture now stabilizes at $W = 8.75$ cm (red curve in Figure 12.12a) as the precipitation minus evaporation again reaches zero (Figure 12.12b). In this case, as the prescribed precipitation is lower, the equilibrium evaporation must also be lower. This is achieved again via the term in the model equation representing the resistance to evaporation, W/W_k, which results in a lower evaporation due to the lower soil moisture content W. The conclusion is that as precipitation changes, the evaporation rate adjusts in order to achieve a new equilibrium. The new equilibrium requires some drying of the soil, so that the evaporation rate equilibrates with the lower precipitation rate. The formulation of the resistance to evaporation in more complex models typically represents a more nonlinear dependence on the soil moisture W, allowing a large response of soil moisture to a small change in precipitation under some circumstances. In reality the precipitation may also depend on the local evaporation, making the dynamics even richer. The existence of surface runoff and groundwater flow that transport water between regions implies that in a more realistic scenario the evaporation minus precipitation does not necessarily vanish at a steady state but balances such transports instead.

Having gone through the way soil moisture is calculated from precipitation in this bucket model, we can better understand the projections examined in section 12.5 for the Sahel and Southwest United States. Note that the difference in precipitation over the Sahel between the two models seen in Figure 12.5a is not large, but the difference in projected soil moisture seen in Figure 12.5c is greatly amplified. Similarly, the slightly larger net evaporation predicted by the MIROC model in the Southwest (Figure 12.5e) leads to a large decrease in projected soil moisture there. In both cases, soil moisture integrates over longer-term forcing trends. The dependence of the soil resistance to evaporation on soil moisture (that is, the factor W/W_k in the bucket model equation 12.5, or a more nonlinear dependence in more sophisticated formulations) leads to a potentially sensitive response of soil moisture to meteorological forcing: a small change to precipitation can lead to a large response of soil moisture.

12.8 WORKSHOP

A Jupyter notebook with the workshop and corresponding data file are available; see https://press.princeton.edu/global-warming-science.

1. **Past and future of Sahel and Southwest United States droughts:**
   Plot the provided results of the two climate models for 1850–2100, based on a projected RCP8.5 scenario, showing the following time series for the Sahel: (1) evaporation and precipitation (in m/yr), (2) evaporation minus precipitation, (3) soil moisture. Repeat, in three more panels, for the Southwest United States. Describe what you see in each panel, discussing the likelihood of future droughts in each region and the expected uncertainty.

2. **Identifying ACC in a long-term Southwest United States drought record:**

   (a) Plot the 2000 yr time series of PDSI from the North American Drought Atlas.

   (b) Use non-parametric statistics to figure out if the number of droughts in the most recent $N = 30$ yr in the record is unusual within 95%, defining a drought as being PDSI < -1.5, and using 10,000 realizations.

3. **Tree-ring data:** Plot (in thin gray lines) the time series of tree-ring width records for bigcone Douglas-fir trees from the San Bernardino Mountains as a function of time. Plot (in a thick, color line) the decadal bin-average. Comment on the scatter around the average and discuss the implications for the uncertainty of the reconstructed drought record based on these data.

4. **"Wet getting wetter, dry getting drier" projections:**

   (a) Plot the zonally averaged precipitation minus evaporation as a function of latitude for 1920–1940 and for 2080–2100 under RCP8.5.

   (b) Plot the projected change in zonally averaged precipitation minus evaporation as a function of latitude, and superimpose the estimated expected change based on the Clausius-Clapeyron scaling in equation (12.2).

5. **Extreme precipitation events:**

   (a) Plot the PDF of daily precipitation at 40°N for 1920–1940 and for 2080–2100 under RCP8.5. Calculate the 99.9th percentile precipitation rate (mm/day) for each period.

(b) Plot the 99.9th percentile precipitation rate as a function of latitude for the two periods, superimposing the corresponding smoothed curves.

(c) Plot the projected percent change per degree Kelvin warming of the 99.9th percentile precipitation rate as a function of latitude based on the above smoothed curves. Superimpose the expected percent change in extreme precipitation per degree Kelvin based on the expected change to the surface saturation specific humidity (that is, based on the Clausius-Clapeyron scaling).

(d) ***Optional extra credit:*** Plot the Clausius-Clapeyron scaling for the expected increase in extreme precipitation rates with temperature, based on the dependence of the saturation specific humidity on temperature T, from 230 K to 320 K. Superimpose the dependence of the thermodynamical component of equation 12.3, normalized to have the same value as the Clausius-Clapeyron scaling at $T = 230$ K. Discuss the difference between the two estimates. Why are these estimates less relevant in the tropics?

6. **Bucket model for soil moisture:**

(a) Run the bucket model with a precipitation rate of $P_1 = 2.2$ m/yr and then $P_2 = 1.8$ m/yr. Plot $W(t)$ and $P - E$ as a function of time. Why doesn't the soil completely dry when the precipitation rate is decreased from the value P_1 that led to the first equilibrium to the lower value P_2? Describe and explain the stages seen in the solution time series.

(b) ***Optional extra credit:*** Drive the bucket model with a precipitation rate of 2 m/yr plus a white noise time series with a zero mean and a specified std. Compare the occurrence of meteorological versus hydrological drought events from the model output via their PDFs and power spectra, and rationalize your results.

7. ***Optional extra credit:*** **Composite analysis of drought years:**

(a) Analyze the provided results from a long control run of a climate model to calculate and contour the rainy season averages (composites) of sea level pressure and precipitation anomalies for times in which the area-mean precipitation over the region of your city of birth is 1 std below its long-term average, and when it is 1 std above the average. Discuss your results.

(b) Plot, describe, and explain the precipitation over the tropical Pacific, from Australia to South America, averaged over times corresponding to El Niño versus La Niña events. Define the

times of these events as the NINO3.4 index being above and below 1 std.

8. **Guiding questions to be addressed in your report:**

(a) Why are droughts important to understand and predict?

(b) Why and how can they change due to natural climate variability? Due to anthropogenic climate change?

(c) How does a deficit in precipitation affect soil moisture? What are the roles of the duration and magnitude of such a deficit?

(d) How and based on what data can we determine if a given drought may be attributable to anthropogenic climate change? What are the uncertainties in such a determination?

(e) Discuss projected drought conditions for the Southwest United States and for the Sahel and their expected socioeconomic consequences.

(f) What can we say about the projection of global-scale precipitation changes in a warmer climate? What about regional changes?

(g) What are extreme precipitation events, why are they expected to increase, and what are some socioeconomic consequences?

13

HEAT WAVES

Key concepts

- Heat waves as weather events, location-specific threshold temperature and duration
- Processes: high pressure aloft, subsidence, surface winds, clear sky and enhanced shortwave radiation, dry soil
- Heat stress and human health effects
- Projections: anticipated changes to amplitude, frequency, duration, and total number of heat wave days
- Understanding the projected shift in heat wave statistics

A heat wave is a period of unusually hot weather lasting at least a few days. The definition is relative to normal conditions, and therefore the thresholds of temperature and duration vary from region to region. The thresholds may

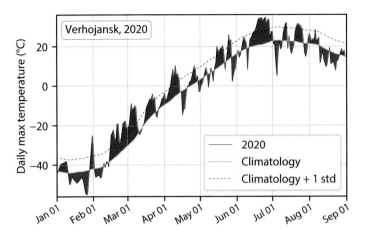

Figure 13.1: The Siberian heat wave of 2020.
Daily maximum temperature in Verhojansk, Siberia (latitude 67.6°N, longitude 133.4°E) during part of 2020, showing the summer heat wave there. The dashed line shows the climatology plus one standard deviation.

be absolute (e.g., maximum daily temperature larger than 40 °C and duration longer than three days) or relative (e.g., temperature and/or duration above the 95th percentile). Heat waves are extreme weather events, and severe ones are categorized as natural disasters. They can lead to the failure of agricultural crops, power outages due to increased use of air conditioning, and effects on human health, including death from heat stress and overheating. Enhanced frequency and amplitude of heat waves can also affect natural ecosystems and wildlife, leading, according to one study, to higher water requirements and therefore to increased mortality of desert birds. Not surprisingly, heat waves are predicted to become more frequent, stronger, and longer in a warmer climate. Because these events last a few days to a few weeks, they differ from droughts, which are longer by definition, although they do share several common elements, as discussed below.

A significant increase in the occurrence of heat waves is already clearly observed in many locations around the globe. An especially impressive heat wave event occurred in Verhojansk, Siberia, north of the Arctic circle, during 2020, as shown by the time series of the maximum daily temperature in Figure 13.1. The heat wave led to a temperature of about 34 °C around June 21, some 13 °C above the climatological mean, and the temperature was more than one standard deviation above the mean for over two weeks. This heat wave, which followed unusually warm weather during the preceding months, occurred with widespread fires, which are believed to have also caused significant CO_2 emission, permafrost melting that led to infrastructure failure, fuel leakage to a river and an environmental catastrophe, and significant health impacts.

In section 13.1, we review the physical processes that lead to extreme heat waves, demonstrate them using a test case of heat waves over the Great Plains in North America, and discuss their similarities, differences, and relation to drought events. The roles of relative humidity and the *wet bulb temperature* that can be used to quantify the health danger of heat stress conditions are discussed in section 13.2. We then analyze model projections of increased heat wave amplitude, duration, and number in a warmer future climate and demonstrate how the change in these statistics may be understood (section 13.3).

13.1 PHYSICAL PROCESSES

The extreme and persistent warming during a heat wave is due to the combination of several factors discussed in this section. Figure 13.2 shows the analysis of heat waves occurring over the Great Plains in North America during 1920–1970, as simulated by an ensemble of some 30 climate model runs. The daily maximum temperature (the maximum temperature that occurs during a given day, T_{max}) anomaly pattern (calculated as the deviation from the mean August daily maximum temperature), averaged over many extreme heat wave events, is in Figure 13.2a, showing temperatures that are up to $10\,°C$ above normal summer daily maximum temperatures for the corresponding period.

Heat waves typically occur under a persistent high pressure region aloft, in the mid-troposphere at a height of 3–7 km. Commonly, this is presented via the height of the 500 hPa pressure level (color shading in Figure 13.2c). This is referred to as the 500 hPa *geopotential height* and denoted $z(500\,\mathrm{hPa})$, and a region with a higher geopotential height implies high pressure there relative to the surrounding regions. The high pressure region diverts the high level winds to flow around it (vectors in Figure 13.2c). This wind pattern corresponds to an anomalous clockwise circulation around the high pressure (vectors in Figure 13.2d). The flow around the high pressure is due to the Coriolis force, as discussed for droughts (section 12.2) and for the reverse case of the flow around

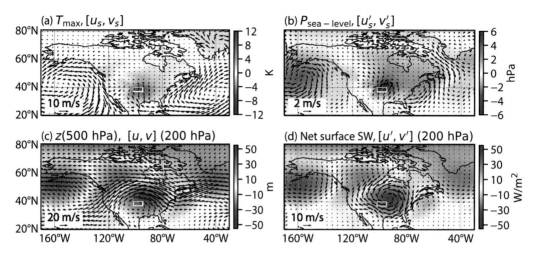

Figure 13.2: Heat wave mechanisms.
Analysis of heat waves simulated in a climate model over the central plains of North America for latitudes 37°N–40°N and longitudes 95°W–102°W, marked by white rectangles. Shown are averaged fields over periods corresponding to heat waves (that is, heat wave composites), where a heat wave is defined as the daily maximum temperature being above 39 °C for at least 3 days. (a) Maximum daily surface temperature (T_{max}) anomaly (calculated here as the deviation from the August mean) and surface winds. (b) Sea level pressure anomaly and surface wind anomalies. (c) Geopotential height anomaly at 500 hPa and winds at 200 hPa. (d) Net surface shortwave radiation anomaly and anomaly winds at 200 hPa. Vectors in panels a and c show the surface and 200 hPa wind fields averaged over heat wave events, while panels b and d show the deviation of these wind fields from their August means.

a low pressure in the context of hurricanes (section 8.1). The diversion of the large-scale upper atmospheric wind by the high pressure center aloft prevents precipitating storms from reaching the area under the high pressure, keeping the soil dry. Dry soil exacerbates heat waves, as it does not allow the ground to cool via the evaporation of soil moisture (latent heat flux). This means that a precipitation deficit during the recent rainy season can also lead to more/stronger heat waves during the warm season.

The high pressure aloft also leads to subsidence (sinking motion) and therefore to an adiabatic warming of the sinking air as it encounters higher pressure levels while descending, and therefore warms, increasing the surface temperature and thus the heat wave amplitude. This adiabatic compression and heating can also be understood by considering the dry static energy of subsiding air parcels. As a reminder (section 7.2), *dry static energy* (DSE) is the sum of the internal energy of a parcel due to its temperature and its potential energy due to its elevation, and is defined as $DSE = c_p T + gz$, where T is the air temperature, c_p is the specific heat capacity at constant pressure, g is gravitational acceleration, and z is the elevation in meters. DSE is conserved following air parcels undergoing adiabatic motions (no mixing and no heat exchange with the surrounding air) in the absence of condensation and evaporation. As the parcel descends, its potential energy (gz) decreases, and therefore its internal energy ($c_p T$) increases to compensate and conserve the total energy, hence the warming. The warming of the subsiding air, in turn, increases its saturation specific humidity according to the Clausius-Clapeyron relation (Box 2.1) and therefore lowers its relative humidity (as is the case in droughts; see section 12.2). This drying prevents atmospheric convection and cloud formation (section 7.2) and leads to a clear sky. This, in turn, causes a significant increase in shortwave radiation at the surface and to further heating there (Figure 13.2d).

The subsidence in a heat wave is strengthened by the clear sky it induces: clouds absorb LW radiation emitted by the atmosphere below and emit it back toward the cooling air. The lack of clouds allows the subsiding air within the heat wave to emit LW radiation to outer space and cool more effectively. This effective radiative cooling causes the air to become denser and to sink to lower altitudes: such longwave radiative cooling of mid-tropospheric air typically occurs at a rate of 1 K/day, leading to a large-scale averaged subsidence of about 100 to 200 m/day. This subsidence rate is enhanced under the clear sky conditions of a heat wave event.

Adiabatic compression of subsiding air also plays an important role during the formation of heat waves. Because DSE is conserved following air parcels undergoing adiabatic motions, air parcels in the mid-troposphere away from the atmospheric boundary layer (lower 1–2 km) tend to travel along surfaces of constant dry static energy. During the development of heat waves, air parcels that eventually end up in the heat wave region are transported by winds along downward-sloping DSE surfaces. These parcels lose elevation (gz decreases) and experience compression and therefore adiabatic heating ($c_p T$ increases to compensate), thus drying via the reduction of their relative humidity. Studies of the formation of specific heat wave events show that the warming due to adiabatic compression is a dominant factor in leading to the hot conditions and is much larger than the radiative cooling that accompanies the subsidence.

A low pressure pattern that sometimes develops at the surface near the center of the heat wave (known as *heat low* or *thermal low*; see Figure 13.2b), while sometimes less obvious and distinct than the upper level high pressure pattern, leads to surface winds from warmer lower latitudes, again enhancing the warming. This is seen in Figures 13.2a,b, which show the full surface winds during heat waves and their anomaly from the August climatology, correspondingly. In the case of heat waves in other regions, the surface winds may blow from high to low elevation, leading to further enhanced subsidence and warming, or from warm continental interiors toward coastal areas, leading to heat waves there.

The high pressure aloft during heat waves tends to remain fairly stationary, typically for a week or two, unlike typical weather high and low pressure centers that tend to move continuously. As such, the stationary high pressure is sometime referred to as a *blocking event*, and its persistence leads to the extended length of the corresponding heat wave. A final factor enhancing the amplitude of a heat wave is that the atmosphere, having undergone drying due to the subsidence, does not support moist convection (section 7.2) and the corresponding intense vertical motions (updrafts and downdrafts) and therefore does not mix air vertically effectively. As a result, the high horizontal wind velocities that typically exist aloft are not communicated to the surface, and the surface wind is therefore weaker. This implies less surface evaporation (e.g., section 12.7) and therefore less surface cooling via latent heat flux.

To summarize, a heat wave is caused by a persistent high pressure aloft, which leads to subsidence and thus to a warming of the sinking air. The drying of the air due to the subsidence and warming lowers its relative humidity and leads to cloud-free, clear sky conditions, which allow more shortwave radiation to reach the surface. The clear sky also enhances the radiative cooling of air and strengthens the subsidence. The high pressure diverts precipitating storms, which prevents soil moistening and the associated latent heat cooling. Lack of precipitation in the previous rainy season and the resulting drier soil act as a preconditioning that can enhance the occurrence of heat waves by again reducing cooling via surface evaporation and latent heat flux. Surface winds, directed in the specific example we examined by a low surface pressure pattern, further enhance the warming by bringing warm air toward the region affected by the heat wave. Depending on the heat wave location, the surface winds may be from warmer lower latitudes, from a high to low elevation enhancing warming due to subsidence, or from a warm continental interior toward a coastal area where the heat wave is occurring. Finally, the dry atmosphere in the heat wave region prevents atmospheric convection and therefore prevents mixing of the surface air with the upper atmosphere, where winds are typically strong. This lowers the surface winds and as a result weakens surface evaporation that can cool the surface.

Many of these factors leading to heat waves are also common to droughts (chapter 12). Droughts are longer-term events that require some longer-term sustained forcing, such as a remote persistent SST pattern that drives an atmospheric Rossby wave train, inducing a high pressure center over the drought region. The persistent remote forcing allows droughts to be maintained for long periods. Heat waves, on the other hand, are weather events mostly lasting a few days to a couple of weeks and generally occurring due to a random combination of related weather factors and feedbacks rather than due to longer-term persistent external forcing. The longer duration of droughts allows them to be reinforced by affecting their own forcing. For example, drought conditions lead to soil drying, which further increases the temperature anomaly and strengthens the drought. Heat waves, while not requiring sustained remote forcing, may be strengthened by various longer-term preconditioning such as a drier prior rainy season or dry soil in general. Given the many common factors leading to and strengthening both droughts and heat waves, it should be clear that longer-term drought conditions may lead to more, stronger, or longer heat waves. This was likely the case, for example, for the 1936 North American heat wave, which took place during the Dust Bowl period of sustained droughts during the 1930s.

13.2 HEAT STRESS

Heat waves can impact human health by creating a heat stress physiological condition, in which the body is unable to maintain a healthy temperature in a hot environment. This occurs depending not only on the temperature of the environment but also on the humidity, because at a higher relative humidity the body has difficulties cooling by evaporating sweat. A simple measure of this effect is the *wet bulb temperature* (WBT), defined as the temperature that an air parcel would have if cooled adiabatically at a constant pressure until its RH is 100% by evaporation of water into it, where all the latent heat required for the evaporation is supplied by the parcel. If the relative humidity of an air parcel is 100%, its temperature is equal to its WBT. If the relative humidity of an air parcel is less than 100% and it is exposed to liquid water, some of its heat is used to supply the needed latent heat of evaporation, and the air parcel cools, implying that its WBT is lower than its actual temperature. The WBT in the current climate does not exceed 31 °C; humans and other mammals have difficulties surviving under long-term exposure to a WBT of 35 °C and above.

The WBT may be calculated using an energy conservation argument for an air parcel. For this purpose we quantify the energy of an air parcel by including its internal energy due to its temperature, plus the energy available in the form of water vapor that can condense and release the latent heat of condensation, similarly to the moist static energy argument in the context of atmospheric convection (section 7.2). Consider an air parcel of temperature T and energy $c_p T + Lq = c_p T + L \cdot RH \cdot q^*(T)$, where q is its specific humidity, RH the relative humidity, L the latent heat of condensation, and $q^*(T)$ its saturation specific humidity. After the evaporation of water into the parcel until it is saturated, the parcel is at its wet bulb temperature T_w and at a saturated moisture level of $q^*(T_w)$, and its energy is $c_p T_w + Lq^*(T_w)$. The energy conservation statement is therefore $c_p T + L \cdot RH \cdot q^*(T) = c_p T_w + Lq^*(T_w)$. This is a nonlinear equation for T_w that may be solved graphically or using a root finder optimization routine, as discussed in section 7.2. The solution for T_w is contoured as a function of relative humidity and temperature in Figure 13.3. As an example, a WBT of 35 °C may be obtained either at a temperature of 35 °C and relative humidity of 100% or at a temperature of 50 °C and a relative humidity of 35%.

As mentioned, heat waves involve subsidence aloft and thus adiabatic heating, which can lead to a temperature *inversion*, in which the subsiding air is warmer than the surface air below. This is referred to as an inversion because normally the atmospheric temperature becomes colder with height, of course. When such an inversion develops during a heat wave, the warmer air above the surface is significantly lighter than the surface air, preventing the surface air from rising and mixing with the warmer, lighter, and drier air aloft. This can trap moisture due to surface evaporation near the surface and increase the surface relative humidity. This increase, in turn, leads to a higher WBT and increases the likelihood of heat stress. While we focus in this chapter on the temperature signal of

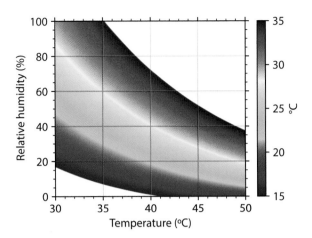

Figure 13.3: Quantifying heat stress risk.
The wet bulb temperature ($^\circ$C) as a function of temperature and relative humidity.

heat waves, it is good to keep in mind that the relative humidity signal could be equally or even more important in terms of impacts on human health.

13.3 FUTURE PROJECTIONS

It is no surprise that a warming climate leads to more, stronger, and longer heat waves, as more warm days are above the threshold amplitude and duration used to define a heat wave in a given region. Figure 13.4 shows results from an ensemble of 30 climate model runs, carried out from 1920 to 2100 using observed forcing up to 2005, followed by the RCP8.5 warming scenario to year 2100. From very few cases of going over a threshold of 40 $^\circ$C in the 1920s, the simulations show summer temperatures persistently above this threshold toward the end of the 21st century. The blue curve shows the average over the 30 runs (i.e., the ensemble mean), and toward the end of the simulation even the mean occasionally crosses the threshold.

One may calculate statistics characterizing heat waves over any given decade, including their mean duration, mean amplitude, the averaged number of events per year, and the total number of heat wave days per year. All of these measures show a very significant increase during the 21st century according to the RCP8.5 scenario, as in Figure 13.5, consistent with the maximum daily temperature time series in Figure 13.4.

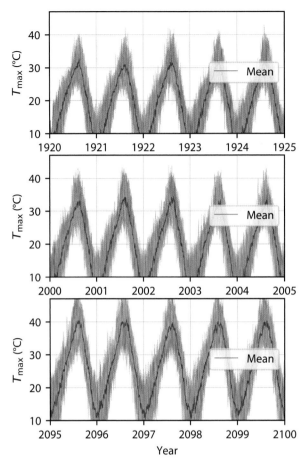

Figure 13.4: Evolution of maximum daily temperatures in a warming scenario.
Maximum daily surface temperature over the Great Plains region marked by white rectangles
in Figure 13.2. Shown are the results of 30 ensemble climate model runs from 1920 to 2100,
using historical observed greenhouse gas forcing until 2005, followed by the RCP8.5 scenario
until 2100. The three panels show time series of daily maximum temperature during three 5-year
periods starting in 1920, 2000, and 2095. The ensemble mean is shown in blue. Note the very
few cases over a threshold of 40 °C in the 1920s vs persistent summer temperatures above this
threshold toward the end of the 21st century.

The probability distribution function (PDF) of the maximum daily temperature in Figure 13.6a represents the probability of encountering a given value of
$T_{\max}$. That is, $PDF(T)\Delta T$ is the probability of finding a maximum daily temperature in the range of $(T, T + \Delta T)$. It shows a shift of the PDF to the right
in the latter parts of the 21st century relative to the beginning of the 20th century (compare blue to red bars), corresponding to many more occurrences of

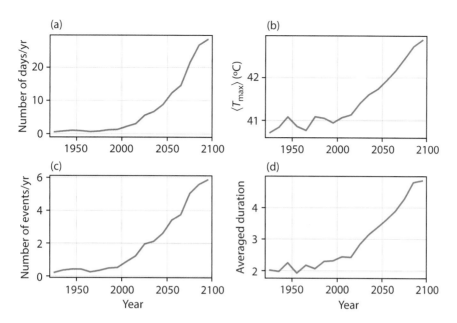

Figure 13.5: Projected heat wave statistics.
Time series of decadal statistics of heat waves over the Great Plains area analyzed in Figure 13.2 from 1920 to 2100. (a) Averaged number of heat wave days per year. (b) Averaged maximum daily temperature in a heat wave. (c) Averaged number of heat waves per year. (d) Averaged heat wave duration in days.

high daily maximum temperatures. Similarly, Figure 13.6b shows that heatwave events become longer, with a longer tail suggesting many events with a duration of up to 15 days. Finally, Figure 13.6c shows that the heat wave events are characterized by a larger averaged amplitude in a projected warm climate than at the beginning of the 20th century. The particular numerical values for the statistics of duration, amplitude, and frequency of heat wave events depend on the threshold parameters used to define heat waves, but the overall trend of increasing duration and amplitude in a warmer climate is, of course, very robust.

The shifts in the characteristics of heat waves could, in principle, be due to two possible effects. First, the increase in the mean temperature with warming means that extreme thresholds are crossed more frequently. Second, there may be a shift in higher order moments of the statistics of the events. For example, the mean of T_{max} could shift and the variance could increase, each leading to more extreme events in a warmer climate. To test which of these two processes is more dominant, we take a daily maximum temperature time series from 1920 to

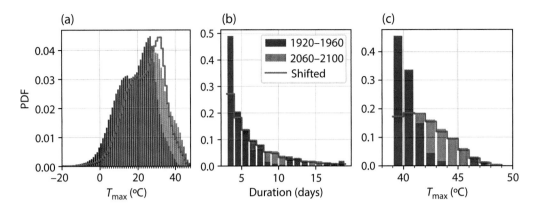

Figure 13.6: Understanding changes in heat wave statistics.
Heat wave statistics during the early 20th century vs the end of the 21st century in the RCP8.5 scenario. (a) Probability distribution function for the simulated daily maximum temperature for the period 1920–1960 over the Great Plains region marked by white rectangles in Figure 13.2 and for 2060–2100 in an RCP8.5 warming scenario. (b) Probability distribution function for heat wave duration and (c) for average daily maximum temperature during heat wave events. Shown are the PDFs for the simulated period of 1920–1960 in blue and for the projected period of 2060–2100 in red. The green lines represent the PDFs calculated from the 1920–1960 time series of daily maximum temperatures, with an added constant temperature equal to the mean difference between the periods of 1920–1960 and 2060–2100.

1960, add a fixed constant temperature shift corresponding to the mean increase in T_{max} in 2060–2100, and calculate again the different PDFs (green lines in Figure 13.6). The distributions of daily maximum temperature, duration, and mean heat wave amplitudes calculated by the shifted time series (green line) are nearly identical to those calculated from the maximum daily temperatures from the 2060–2100 time series (red bars).

The similarity of the heat wave statistics for the latter part of the 21st century to those deduced from the early 20th-century temperatures plus a constant shift suggests that the physics of heat wave events does not change in this warmer climate projection. Instead, it is simply the shift in the mean daily maximum temperature that leads to the pattern of increased number, duration, and amplitude of heat wave events. Of course, this assumes that the model simulation of future heat waves used for this analysis is reliable. Because heat waves are weather events, it is difficult to attribute a given event to ACC. A detection of climate change in observed heat waves in a specific location requires a record that is sufficiently long to include many such events. Then, plotting probability distribution functions of several characteristics, similar to those in Figure 13.6, we can

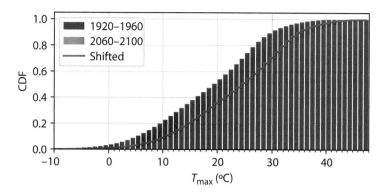

Figure 13.7: Increased probability of exceeding a given maximum daily temperature in a warm climate.
The cumulative probability distribution function for the simulated daily maximum temperatures for the period 1920–1960 (blue), and for 2060–2100 in an RCP8.5 warming scenario (red) over the Great Plains region marked by white rectangles in Figure 13.2. The CDF shown by the green line is calculated as in Figure 13.6.

compare to the earlier record during the 20th century in order to detect a change in the heat wave statistics reflecting the effect of climate change.

The probability of exceeding a given daily maximum temperature, or a given duration or amplitude of a heat wave, can be estimated from the corresponding cumulative distribution function (CDF), which is the integral over the PDF, $\mathrm{CDF}(T) = \int_{-\infty}^{T} \mathrm{PDF}(T')dT'$. The CDFs for the daily maximum temperature for the early 20th century and late 21st century are in Figure 13.7. The value of the CDF at a temperature $T_{\max}$ is the probability of finding a temperature *below* that value. One minus that value therefore provides the probability of finding temperatures *above* that value. From the figure we deduce that the probability of exceeding, say, 40 °C during the entire year was 0.1% early in the 20th century and is expected to be about 5% late in the 21st century, according to the RCP8.5 scenario. These probabilities would be higher, of course, if only the summer months were considered, but the expected increase in probability of occurrence of warm temperature is robust and very significant.

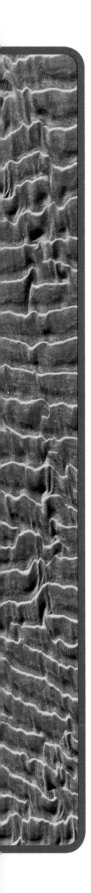

13.4 WORKSHOP

A Jupyter notebook with the workshop and corresponding data file are available; see https://press.princeton.edu/global-warming-science.

1. **The Siberian heat wave of summer 2020:**

   (a) Plot the daily maximum temperature climatology for Verhojansk, Siberia, from January to December, and then superimpose the 2020 temperature.

   (b) Plot the anomaly time series, calculated as the deviation from the daily climatology, and use it to calculate the peak time, amplitude, and length of the summer heat wave based on what you think are reasonable thresholds.

2. **Heat stress and the wet bulb temperature:**

   (a) Calculate the wet bulb temperature (to within 0.001 °C) for an air parcel with temperature of 37 °C and relative humidity of 60%: evaluate the appropriate energy balance for different values of T_w until you find the one that satisfies it.

   (b) *Optional extra credit:* Calculate and contour the WBT as a function of relative humidity and temperature.

3. **Understanding heat wave statistics for an RCP8.5 projection:**

   (a) Plot simulated temperature time series for all ensemble members for 5 yr periods starting in 1920, 2010, and 2090.

   (b) Calculate and plot PDFs of maximum daily surface temperatures for 1920–1950 (blue bars) and for 2070–2100 (red bars). Superimpose a line for the PDF calculated from the time series of 1920–1950 plus a constant temperature increase corresponding to the mean difference between the two periods.

   (c) Calculate and plot the CDFs for the daily maximum temperature for 1920–1950 and for 2070–2100. Superimpose a line for the CDF calculated from the time series of 1920–1950 plus a constant temperature increase corresponding to the mean difference between the two periods. Use the CDFs to calculate the expected probability of exceeding a daily maximum temperature in the latter period, whose probability was 0.1% in the earlier period.

   (d) *Optional extra credit:* Calculate and plot the PDFs of daily maximum temperature, heat wave duration, and heat wave amplitudes following the above guidelines in item (b).

4. **Heat wave composite analysis in the region of your city of birth:**

   (a) Calculate and plot a climatology of August daily maximum temperature and of Z500 (geopotential height of the 500 hPa surface).

   (b) Calculate a time series of maximum daily temperature over the city of interest (averaging over an area of about 5 × 5 degrees).

   (c) Use the time series to calculate and plot composites for heat wave events of both fields plotted in (a). Define a heat wave as having temperature of more than an appropriate threshold, ignoring the duration condition for simplicity.

5. **Decadal heat wave statistics:**

   (a) Calculate and plot decadal time series of the number of heat wave days over the Great Plains from 1920 to 2100.

   (b) **Optional extra credit:** Calculate and plot the decadal statistics of the averaged amplitude, duration, and frequency (number of events per year) of heat waves for 1920–2100.

6. **Guiding questions to be addressed in your report:**

   (a) Why do heat waves occur in the present climate?

   (b) What are the different factors playing a role in their mechanism?

   (c) How is the character of these events expected to change in a warmer future climate? Why?

   (d) What are the implications for human health, agriculture, other activities? In your answer, consider separately regions that are currently classified as having temperate climates versus regions that are currently deserts or semi-arid climates, according to the *Köppen climate classification*.

   (e) Given that heat waves are defined to be extreme and therefore rare events, what are the expected difficulties in attributing individual such events to anthropogenic climate change? Why is it still easier to attribute heat waves to ACC than drought events?

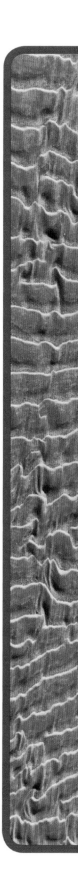

FOREST FIRES

Key concepts

- Fuel aridity and fire weather
- Non-climate-related human influences: ignition, fire suppression management and fuel accumulation, population increase
- Climate factors: droughts, heat waves, temperature, prior rainy season precipitation, winds, vapor pressure deficit
- Fires enhanced by climate variability versus anthropogenic climate change
- Test cases: western United States and Australia
- A global perspective and future projections

Fires in Australia and the western North America, to name two examples, seem to have been especially dramatic in recent years, repeatedly breaking records and raising the possibility of a connection to anthropogenic global warming. Forest fires are a danger to humans and domestic and wild animals, and they destroy property. They emit noxious aerosols and lead to the formation of toxic ozone levels, as well as of PM2.5 particles that are known to lead to health problems. The smoke spreads over large distances, disrupting life hundreds to thousands of kilometers away from the fire. Fires also contribute significantly to CO_2 emissions, and it is estimated that they contributed 1.8–3 PgC (10^{15}g carbon) per year over the past decades. This is a significant contribution, even relative to the approximately 10 PgC per year due to the burning of fossil fuel—although much of this emission by fires is natural and what part can be considered anthropogenic is more difficult to estimate. At the same time, lightning-ignited fires, and forest fires specifically, are a naturally recurring, important ecological process. Plants adapt by developing strategies of resistance to fire, post-fire stimulated growth, heat-stimulated germination and flowering, and more. The cones of the lodgepole pine, for example, are sealed by resin and open to release their seeds only after being exposed to high temperatures that occur during fires. The plant/tree species composition of a forest is affected by fire, and plant species and evolution affect fire regimes as well, leading to a complex and evolving two-way coupling. A long record of past fires is preserved as fire scars in tree-ring records and as charcoal in lake sediments, indicating that fires are indeed naturally occurring regular phenomena.

Fires develop and spread due to a combination of three ingredients: fuel, ignition, and dryness, and all three are affected by climate. Climate factors that affect both fire frequency and size include warming trends, droughts, dry weather, higher prior rainy season precipitation that leads to more fuel availability, earlier snow melt during spring, and strong and dry local winds (e.g., the Santa Ana winds in Southern California) that can dry potential fuels as well as enhance and spread fires. Increased lightning activity in some areas due to climate warming can also lead to more ignitions and increased fire activity. On a longer timescale, the climate of a given region affects the vegetation that develops there and thus the potential for fire.

Attributing forest fires to anthropogenic climate change is more complex than is the case in other observed changes we have considered. In the analysis of

Arctic sea ice melting, droughts, hurricanes, and other phenomena, the detection of a signal due to anthropogenic climate change requires identifying a trend that can be separated from natural climate variability. However, in the case of forest fires, human effects include many factors in addition to climate change, so detecting a trend that is not due to climate variability is not sufficient, as the human-caused signal can be non-climate-related. This makes a confident attribution to climate change especially challenging. Among the non-climate-related human effects is the fact that a vast majority of the ignitions of forest fires in some areas are caused by humans. In the United States, human ignitions have accounted for 80% of fires and for about half of the burned area. Such human ignitions also vastly expand the fire season beyond that naturally caused by lightning. Forest fire suppression policies in the western United States, for example, caused an increase in forest density and buildup of fuels, causing what is known as a *fire deficit*. This causes fires to be limited by fuel aridity rather than by fuel availability, and therefore forest fires become more sensitive to the effects of anthropogenic climate warming. Things are further complicated by the fact that the effect of fire suppression depends on the fire regime that existed previously. Forests that have naturally adapted to frequent low-intensity fires (e.g., the western United States) will show higher sensitivity to fire suppression than the wet forests of Alaska and western Canada, where dry conditions are rare and therefore fires are infrequent and intense. The effects of logging, conversion of land to and from agriculture use, and other land-use changes can lead to either increase or decrease in fire activity. Population increases near forested areas lead to more ignitions, and building in areas that are fire-prone induces more fire damage. As the population increases even more, fire suppression increases, and the net effect can be a reduced fire activity. Additional factors that may affect fire trends are changes to the composition of forests (say from one tree type to another or from large to small trees) that are not easily attributed to a specific cause, trends in number of trees per unit area of a forest (forest density), the effect of past fires on the connectivity of forests and thus on the ability of future fires to spread over large areas, as well as the effect of fires on fuel availability in the next few years. The statistical analysis of fire trends and their relation to climate change is made more difficult by the inherently episodic nature of forest fires on timescales from decades to millennia.

Given this complexity, the attempt to identify the effect of anthropogenic climate change on forest fires must resort to a statistical analysis, and we demonstrate here a few example questions that may be attempted. First, how much of the burnt area in the western United States can be attributed to anthropogenic climate change (section 14.2)? Second, what is the effect of climate variability modes on droughts and therefore on bush fires in Australia (section 14.3)? Finally, what can be said about global fire statistics and future projections (section 14.4)?

14.1 TOOLS

The analysis of the effects of anthropogenic climate change on forest fires relies on the identification of climate factors that affect the frequency and magnitude of fire events. Fire is often quantified in terms of the area burnt, and one looks for various climatic and other factors that are well correlated with this measure. Some indices that were found to be predictive of, or correlated with, the area burnt are listed here.

Vapor pressure deficit (VPD): The saturation water vapor pressure ($e^*(T)$; see Box 2.1) minus the actual vapor pressure (e). VPD represents the *moisture demand* of the atmosphere, where a higher VPD means more ability to evaporate water into the air and therefore potentially more rapid drying of fuels. Due to the Clausius-Clapeyron relation, the saturation water vapor, and therefore the VPD, increases exponentially with temperature if the moisture content (say specific humidity) of the air is constant. Furthermore, VPD increases with temperature even if the *relative* humidity (RH) remains constant, although the moisture content increases with temperature at a fixed RH (see this chapter's workshop). The strong correlation of VPD with the burnt area therefore suggests an important role for warming in the occurrence of fires.

Climatic water deficit (CWD): The seasonally integrated potential evapotranspiration (PET; see section 12.1) minus the actual evapotranspiration.

Fire indices: These indices use a variety of factors to predict daily fire danger. Some commonly used fire indices are the Canadian Fire Weather Index (FWI), the Australian (McArthur's) Forest Fire Danger Index (FFDI), and the US Burning Index. They depend on daily weather measurements, including temperature, relative humidity, wind speed, and precipitation. They also take into account factors such as fuel dryness/aridity, fine fuel moisture, drought, the buildup of fuel, and the ability of the fire to spread.

14.2 DETECTION OF BURNT AREA DUE TO ACC

Forest fires in the western United States were especially devastating during the second decade of the 21st century, with previous records of burnt area and socioeconomic damages repeatedly broken. Our objective here is to estimate the forest fire area in the western United States that can be attributed to ACC. Consider first the time series of the $\log_{10}$ of the area burnt in the western United States as a function of year in Figure 14.1a (red curve). One relevant climate factor, the VPD over the same area, is also in Figure 14.1a (blue). The two seem to co-vary and are plotted against one another in Figure 14.1b, together with the corresponding regression line expressing the $\log_{10}$ of area burnt in terms of the normalized VPD,

$$\log_{10}(\text{area burnt}) = a \times VPD + b. \tag{14.1}$$

The deduced relation between area burnt and VPD is robust statistically (p-value smaller than 0.001), and it explains over 75% of the variance. If the two time series (log of area burnt and VPD) are linearly detrended, the regression slope a decreases by only about 10%, and the explained variance decreases only very slightly to 72%, a surprisingly strong result. This strong and robust statistical relation is especially remarkable because large fires depend on so many other factors beyond the dryness of fuel that is induced by high VPD values: they require an ignition, strong winds, fuel availability, and the ability to grow in spite of human attempts to extinguish them (97% of fires in this region are eliminated by humans before becoming large enough to significantly affect the total annual burned area).

This regression can now be used to estimate the contribution of ACC to the area burnt. For this purpose, we estimate the ACC signal in VPD separately from natural variability. This is done by considering VPD time series from many climate model projections, averaging them together and further smoothing them in time to eliminate the signal due to natural climate variability. Figure 14.2 demonstrates the extraction of an ACC signal (in temperature rather than VPD in this case) from model-estimated natural variability. The figure shows individual time

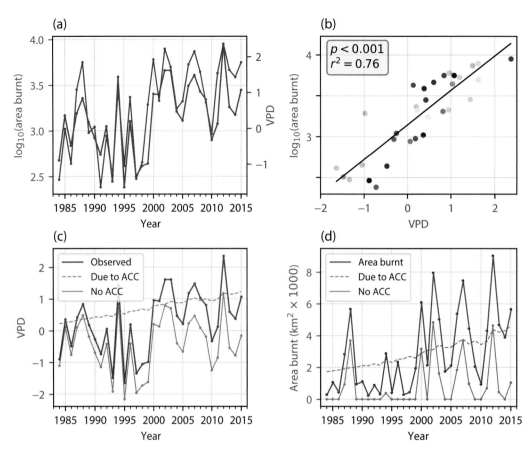

Figure 14.1: Estimating the contribution of ACC to western United States forest fire area.
(a) Red line denotes the $\log_{10}$ of the area burnt in 10^3 km^2. Blue line shows the VPD time series over the same area (in units of standard deviations from the 1979–2015 mean of the VPD time series, after the ACC trends in temperature and specific humidity have been removed). (b) The regression between the $\log_{10}$ of the burnt area and the standardized VPD record. Points are color-coded by year of data, from blue to red. (c) The blue line again shows the standardized VPD time series over the western United States. The dashed line shows the contribution of ACC to the VPD estimated via a multi-model average. The solid gray line is the VPD time series with the ACC signal removed. (d) The red line shows the area burnt as a function of year. The dashed gray line shows the estimated contribution to the burnt area due to ACC calculated from the VPD due to ACC using the regression relation. The solid gray line shows the estimated burnt area without the contribution of ACC.

series of August temperature over the western United States from over 30 model ensemble members (thin color lines), differing only very slightly in their initial conditions. Each model then develops a different weather and different natural variability (e.g., El Niño events occur in different years in each model run). The average over all ensemble members, which leads to the elimination of much of the natural variability signal that is different and out of phase in different runs, is shown by the red curve. The smoothed mean then (thick yellow curve) represents a model-estimate of ACC with the signal of natural climate variability removed.

A similar procedure of averaging over multiple climate model projections and smoothing can be used to estimate the VPD trend due to ACC over the western United States; the result is shown by the gray dashed line in Figure 14.1c. The regression relation (14.1) can now be used to estimate the area burnt due to ACC from the VPD trend due to ACC (gray dashed line in Figure 14.1d). Removing this component from the time series of burnt area, we can plot the expected burnt area without the effect of ACC (solid gray line in Figure 14.1d). Recent work using several climate indices in addition to VPD to obtain a hopefully more robust estimate suggests that approximately 50% of the increase in western US annual forest fire area observed during 1984–2015 was due to ACC trends.

Of course, the assumption that the ensemble model average represents ACC implies that we trust the model prediction of temperature and VPD and its response to increased greenhouse gases. A similar model-induced uncertainty was encountered in section 9.2, for example, when we assumed that the estimate of natural variability of sea ice using climate models is realistic in order to be able to attribute melting over the past decades to ACC. Interestingly, when attempting to detect ACC in observed sea ice melting trends, we relied on the model variability, while in the case of forest fires we rely on the model simulation of the mean response to ACC. Essentially the question we need to answer for a successful detection of ACC in forest fires trends is how much of the observed trend in VPD (blue line in Figure 14.1a) is due to ACC. The signal shown by the blue line may also be interpreted as variability around one value until year 2000, and then around another, higher value until 2015. This qualitative behavior is seemingly very different from the gradual increase in VPD due to ACC estimated from the models (dashed line in panel c). This difference may simply be because the model estimate is an average over many realizations of the

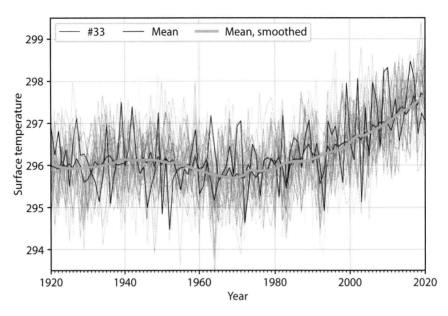

Figure 14.2: Extraction of ACC signal from a multi-model ensemble.
Showing August temperature (K) averaged over the western United States from an ensemble of climate model runs, following the RCP8.5 scenario to year 2020. Thin color lines are individual ensemble members. The black line is ensemble member number 33; the red line is the average over all ensemble members; and the thick yellow line is the smoothed averaged time series representing the ACC signal estimated using this model ensemble.

climate system (being derived from a multi-model average), while the observed line represents a single realization of the actual climate system. The difference between a single realization and the ACC signal is demonstrated in Figure 14.2 as the difference between the thin black line corresponding to a single ensemble member and the thicker yellow curve showing the estimated ACC signal. If the model(s) overestimated the response of VPD to greenhouse gas increase over the period in question (1984–2015), this would lead to an overestimate of the contribution of ACC to the area burnt. A model overestimate of the ACC signal in VPD is but one source of the large uncertainty in the identification of anthropogenic climate change signal in forest fires. Tracking these data over the next decades should allow us to improve our confidence in conclusions regarding the connection between ACC and fires in this area.

It turns out that the frequency of wet days during the fire season also correlates well with the area burnt. Both correlations seem robust, and it is difficult to disentangle which of the two is the prime driver of fires, if any, with the relatively short currently available record. Future model projections of the VPD and of the frequency of wet days are very different. While VPD is projected to increase, the expected mean frequency of wet days is not expected to change dramatically, although its variance may increase. This suggests that the uncertainty with regard to future projections of fire with changing climate is even larger than the uncertainty discussed above based on the role of only one predictor (VPD).

Uncertainty aside for a moment, the fairly tight correlation between climate effects represented by VPD and the $\log_{10}$ of burnt area in the western United States (Fig. 14.1b) means that the burnt area is increasing exponentially with this climate measure. This, if it turns out to be robust, is alarming, of course, as VPD is projected to increase over this area. At some point fires would be self-limiting as a result, but that might occur only after a significant part of the forest area has been damaged. Fires may also cause a permanent shift from a forest to non-forested (e.g., savanna) areas, as in some climates both regimes can exist as alternate stable vegetation covers.

14.3 FIRES AND NATURAL CLIMATE VARIABILITY

Australia experienced a severe drought during 2018–2019 and even earlier, which has likely contributed to the very dramatic fire season from the end of 2019 to the beginning of 2020. While overall climate warming may have contributed to these fires, it seems that natural variability in the Pacific and Indian Oceans was likely a dominant factor leading to a precipitation deficit and persistent drought, and possibly to this intense fire season. El Niño conditions over the East Equatorial Pacific (Box 8.1) have long been known to be a major cause of drought conditions over Australia. Recent work shows, however, that the sea surface temperature over the Indian Ocean, specifically the Indian Ocean dipole (IOD; see Box 12.1) may also play an important role in determining drought conditions over southeast Australia in particular.

The corresponding time series of the IOD and ENSO are in Figures 14.3a,b. Another climate variability mode that is known to affect Australian precipitation is the *Southern Annular Mode* (SAM), represented by the index in Figure 14.3c, calculated as the zonally averaged sea level pressure difference between latitudes 40°S and 65°S. Positive SAM index corresponds to stronger westerlies over

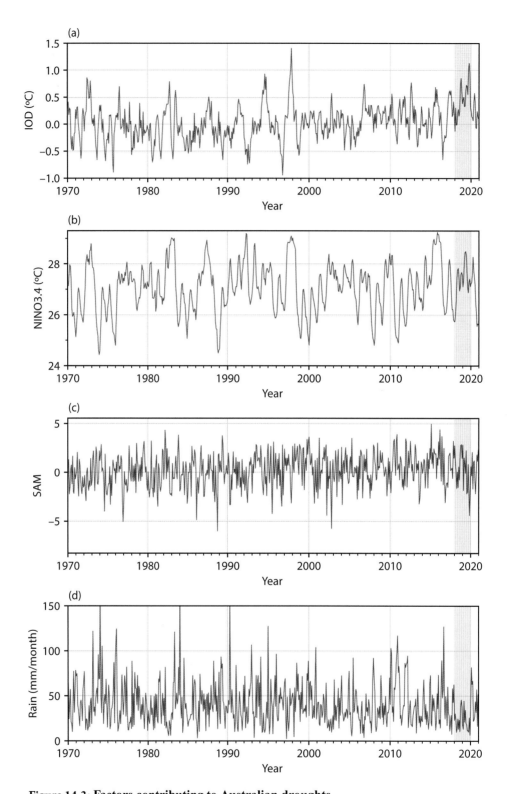

Figure 14.3: Factors contributing to Australian droughts.
(a) IOD index (°C) time series since 1970. (b) NINO3.4 (°C). (c) SAM index (nondimensional). (d) Monthly rain (mm/month) over the Murray-Darling Basin in southeast Australia. Gray shading indicates the dry period of 2018–2019 that preceded the dramatic fire season of late 2019 through the beginning of 2020.

50°S–70°S and weaker westerlies in 30°S–50°S. SAM variability is correlated with many Southern Hemisphere large-scale climate and weather phenomena. Specifically for southeast Australia, SAM influences precipitation by (1) shifting the storm-track and therefore modulating the frequency of precipitating storms arriving to the region and (2) enhancing on-shore wind from the ocean toward southeast Australia, which carries moisture and also modulates precipitation there. These three climate variability modes (IOD, ENSO, SAM) affect droughts (and potentially fires) over large areas, Australia in particular.

The time series of the indices of these variability modes can now be used to explain some of the signal of Australian droughts. Figure 14.3d shows the time series for the rain over the Murray-Darling Basin in southeast Australia, where significant fires occurred during the later part of 2019 through the beginning of 2020. Note the lack of significant rain events over 2018–2019.

Given N monthly values of the IOD, NINO3.4, SAM, and rain over southeast Australia, one can form a $4 \times N$ data set from the time series with their means removed and each normalized by its standard deviation. The N 4-dimensional data vectors can now be clustered (hierarchical clustering was used here) to find the dominant patterns of variability. The clustering groups the points in this 4-dimensional data space that are close together. The cluster analysis shows that the data are best represented by five such groups/clusters.

Figure 14.4 shows the mean location of the monthly data points belonging to each cluster. The blue square represents a cluster of especially rainy months, and as the three panels show, it occurs during a neutral to slightly negative phase of the IOD, during La Niña conditions, and during a positive SAM phase. The average conditions during 2018–2019 (black star in Figure 14.4) clearly show that these years were characterized by conditions that are not conducive to large precipitation: positive IOD index, positive NINO3.4, and neutral SAM. This suggests that the droughts that preceded the dramatic Australian fires of late 2019 through early 2020 may have been the result of a combination of these natural climate variability modes. As should be clear from the analysis, though, the relation between these climate variability modes (ENSO, IOD, SAM) and precipitation over southeast Australia is masked by a larger noise level due to other variability sources, including short-term weather events, and the available observational records are short, all making it difficult to come up with definite conclusions.

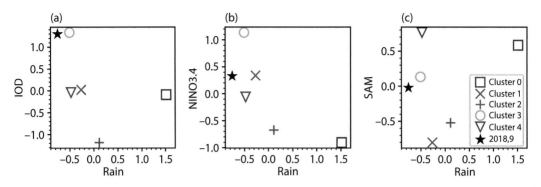

Figure 14.4: Cluster analysis of the Australian drought conditions that preceded the forest fires of late 2019 to early 2020.
Shown are the means for the five calculated clusters. All variables are shown with their long-term means removed and are normalized by their standard deviations. (a) IOD vs rain in the Murray-Darling Basin in southeast Australia. (b) NINO3.4 vs rain. (c) SAM vs rain. The black star shows the mean conditions during 2018–2019.

This analysis does not preclude a contribution of ACC; it is difficult to quantify the contribution of ACC to the state of these three natural climate variability modes during 2018–2019, and an indirect connection between ACC and fires, via an effect of ACC on these modes, remains a possibility. A recent statistical analysis of trends in a specific fire weather index suggests that the probability of the Australian fires during the last months of 2019 and January 2020 was enhanced by anthropogenic warming, with a significant role also played by the natural variability modes examined above.

There are two major challenges that we are facing regarding the interaction of natural variability modes such as ENSO, the IOD, and SAM with fires. First, we need to identify to what degree these variability modes have already been affected by ACC and to what degree events such as the Australian fires of 2019–2020 were influenced as a result. Second, it is important to be able to predict how these modes are expected to change in the future in order to be able to make projections on future changes to fire regimes. While our understanding of the mechanisms behind ENSO, the IOD, and SAM is quite good at present, state-of-the-art climate models still do not simulate them completely realistically and, more importantly, do not always agree on their expected state in a future warmer climate. This makes the prediction of the long-term evolution of these climate modes difficult, making it challenging to produce confident projections of droughts in Australia in a warmer future climate, and thus of fires affected by these droughts. The same holds for fires in the many other areas that are affected by these and similar climate variability modes.

14.4 OBSERVED GLOBAL TRENDS AND FUTURE PROJECTIONS

In addition to the fire indices and other predictors of fires mentioned, one may calculate the length of the potential fire season as in Figure 14.5a. For this purpose, several daily fire indices were calculated and used to define the fire season length at each location. The season length is defined as the number of days during which the fire danger indicated by these indices is above a specified threshold. The global-mean fire season length in the normalized anomaly time series of Figure 14.5a increased by 19% from 1979 to 2013. Similarly, the area affected by the growth in fire weather season length, defined as the total global area where the season length is larger than 1 std above its mean, is in Figure 14.5b. The time series corresponds to an increase of over 100% in the affected area over the same period. The fire season length is found to be correlated with the actual area burnt and with other fire measures in many geographically restricted regions and in many countries. We see that the problem of understanding and predicting fire is complex, though, by the fact that the area burnt globally (Figure 14.5c), actually decreased over the past three decades. This is a result of the many factors affecting fires and fire area, as mentioned in the introduction to this chapter. Specifically, global fire activity is dominated by African fires, and changes in land cover, land use, and population in the African savannalands have reduced the spread of giant lightning-ignited grass fires. At the same time, there is an increasing trend in global closed-canopy forest fires. It is important to understand that a decrease in global fire activity does not mean that warming does not promote fuel aridity. Rather, other trends in fuel moisture, fuel abundance, ignitions, and land use or fire suppression by humans seem to have been more powerful than the forcing by the modest ACC-related warming that has occurred so far, which enhanced fuel aridity and fires.

Given these complexities, it is no wonder that long-term future projections of fire activity involve many uncertainties. One approach to future projections is based on *dynamic vegetation models*, which attempt to simulate within a climate

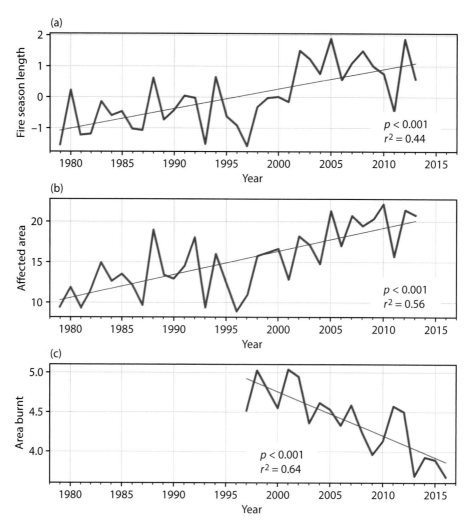

Figure 14.5: Global fire diagnostics.
(a) Global-mean length of the fire season (anomaly, standardized nondimensional measure).
(b) Area affected by longer fire season (anomaly, standardized nondimensional measure).
(c) Estimate of the global burnt area since 1997 (millions of km^2).

model the processes controlling plant growth and changes in vegetation types and their dependence on climate. The simulated vegetation state is then used together with a fire prediction module to predict fires within the climate model. An alternative approach is to use empirical statistical models of fire activity as a function of climate and vegetation types, developed by fitting past fire records, and drive these statistical models with model projections of climate change, including temperature, precipitation, VPD, and others. Unfortunately, discrepancies among climate models, especially when it comes to precipitation changes (section 12.6), can lead to very different fire projections in specific regions.

In addition, future trends in fire activity depend on the current climate regime and vegetation types in a given area. Increased precipitation in currently warm areas characterized by grass and shrubs may lead to higher plant growth and thus increased fires. In moister areas, a precipitation increase can reduce fire activity by increasing fuel moisture. Similarly, warmer and drier weather may increase fires in areas rich with available fuel (e.g., western United States forests), but increases in potential evaporation due to atmospheric warming and drying can reduce the growth of plants in drier areas, thus reducing fire activity. It is therefore important to realize that temperature is not necessarily the dominant factor determining fire activity in a future warmer climate.

Furthermore, there are *indirect* ways in which ACC can affect fires, further complicating the picture. For example, *CO_2 fertilization* refers to the increased rate of photosynthesis and therefore of plant growth due to increased levels of carbon dioxide in the atmosphere, which can lead to increased fire activity. Such fertilization is the dominant, but not only, anthropogenic effect leading to the observed recent *greening* of the Earth, defined as the growth in the leaf area integrated over the growth season. Another often mentioned hypothetical example scenario of an indirect ACC effect is that of bark beetles not dying off in winter due to warmer temperatures, leading to more deadwood and stressed trees, again increasing fire activity. It turns out, though, that the interaction between bark beetle and fire activity is much more complex and depends on the type of fire (e.g., fire spreading through the canopy vs via the surface), time since the beetle attack, moisture and wind levels, and much more.

While even the sign of the response of future fire activity in specific geographic regions is uncertain, it is difficult to imagine a situation where a significantly warmer climate, with its associated changes to atmospheric precipitation and moisture distribution, would not lead to significant changes in fire activity, whether it increases or decreases. The potential for increased fire activity in Arctic and boreal peatlands, which may stop being frozen year-round, or in areas such as western North America seems alarming even if it would be accompanied by reduced fires in other areas. These fire regime changes can be expected to impact human society, natural habitats, and wildlife.

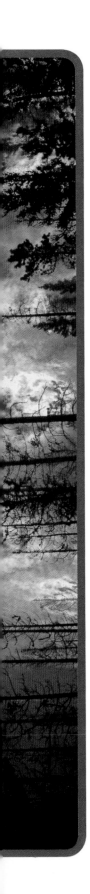

14.5 WORKSHOP

A Jupyter notebook with the workshop and corresponding data file are available; see https://press.princeton.edu/global-warming-science.

1. **Observed fire trends:** Plot the burnt areas in the western United States, Canada, and globally, as well as the number of Canadian forest fires as a function of year. Superimpose a regression line for each of the four plots, calculate the p values, and discuss your findings.

2. **VPD as a fire danger index:**

   (a) Understanding VPD: Plot the saturation water vapor pressure and the water vapor partial pressure assuming an 80% relative humidity versus temperature. Plot the difference between the two curves (i.e., the VPD for a constant relative humidity) versus temperature.

   (b) Plot the western US VPD and area burnt versus time on the same axes. Then repeat using VPD and the $\log_{10}$ of area burnt. Calculate the correlation coefficient between the two plotted time series for each of the two cases.

3. **Separating ACC from variability:** Extract the ACC signal time series in temperature from an ensemble of model runs for 1920–2020. Plot all ensemble time series, and superimpose the ensemble mean and the estimated ACC signal.

4. **Estimate the contribution of ACC to western US forest fire area:**

   (a) Scatter plot the $\log_{10}$ of area burnt versus VPD; calculate and plot a regression line; calculate the r^2.

   (b) Given the VPD increase due to ACC and using your calculated regression, calculate and plot the area that would have burned without ACC; calculate and plot the contribution of ACC as a function of time.

   (c) What percent of area burnt over the last 15 years of data is attributable to ACC?

   (d) Crudely estimate the change in normalized VPD due to a $2\,°C$ warming by examining the provided time series of normalized VPD due to ACC and the western US temperature record in Figure 14.2.

   (e) Given the linear dependence of the log of the area burnt on normalized VPD, how much of the western US forest area (in both 10^3 km^2 and in percent of the total current forest area) might burn per year if the temperature increases by two more degrees Celsius?

5. **Role of variability modes:** Calculate the averaged SAM/NINO3.4/IOD indices over rainy and over dry years in southeast Australia (defined to be above and below one standard deviation, respectively). Compare these averages to the values of these indices averaged over 2018–2019.

6. *Optional extra credit:* **ENSO SST composites:** Calculate El Niño and La Niña SST composites by removing the monthly averaged SST and then averaging over months in which the NINO3.4 time series is 1.5 std above or below its mean.

7. **Guiding questions to be addressed in your report:**

   (a) Describe trends in forest area burnt in western United States and in Canada for the past decades.

   (b) Discuss all *non-climate*-related anthropogenic influences on forest fires and anthropogenic factors that affect fire damage.

   (c) Explain the different ways warming and increases and decreases in precipitation rates may lead to changes in fire trends.

   (d) Discuss the uncertainty sources involved in making future projections of fires on regional and global scales; discuss two specific regions as examples.

   (e) Recommend action apart from an effort to reduce greenhouse emissions.

NOTES

These notes contain references to scientific papers on which the discussion within the book is based and provide details and acknowledgments of the sources of all data used. They also provide a brief mention of a few more refined issues that are beyond the scope of the book itself with relevant references. Each note is preceded by the page number to which it refers.

PREFACE

xi   The graphic element at the top of the title pages, table of contents, and more is based on the global mean surface temperature (upper curve and colors) and on a record of the number of Atlantic hurricanes (lower curve), both plotted from 1880 to 2020. The temperature is from the Had-CRUT4 data set of the UK Met Office Hadley Centre (Morice et al., 2012), and were downloaded from www.metoffice.gov.uk/hadobs/hadcrut4/data/current/ensemble_series_format.html. The record of the number of hurricanes per year is from Vecchi and Knutson (2011), downloaded from www.gfdl.noaa.gov.

xi   Some of the analyses in this book were performed on the high-performance computing resources on Cheyenne (doi:10.5065/D6RX99HX) provided by National Center for Atmospheric Research's (NCAR) Computational and Information Systems Laboratory, sponsored by the National Science Foundation.

xii   While the realism of the RCP8.5 scenario is sometimes doubted, it was recently suggested as potentially plausible to year 2100 (Schwalm et al., 2020). While CO_2 concentration is often mentioned in this book, the CO_2-equivalent mixing ratio is often implicitly implied; this is discussed more explicitly in section 2.2.5.

CHAPTER 1: OVERVIEW

1   A helpful historical perspective of our understanding of global warming is given in the collection of pioneering papers on the subject assembled by Archer and Pierrehumbert (2011).

2   The Mauna Loa CO_2 data used in Figure 1.1a are from Dr. Pieter Tans (NOAA/GML, www.esrl.noaa.gov/gmd/ccgg/trends/) and Dr. Ralph Keeling (Scripps Institution of Oceanography, scrippsco2.ucsd.edu/); see Keeling et al. (1976) and Thoning et al. (1989). The CO_2 record prior to 1958 is a compilation from different ice core records and was downloaded from https://www.epa.gov/climate-indicators.

2   The temperature data for Figure 1.1a are based on the HadCRUT4 data set of the UK Met Office Hadley Centre (Morice et al., 2012); www.metoffice.gov.uk/hadobs/hadcrut4/data/current /ensemble_series_format.html.

2   Figures 1.1b,c are made using output of the CM3 Geophysical Fluid Dynamics Laboratory (GFDL) model (Donner et al., 2011).

2   Data for the blue curve in Figure 1.1d are from Jevrejeva et al. (2008). Data for the red curve are the Global Mean Sea Level Trend from Integrated Multi-Mission Ocean Altimeters TOPEX/Poseidon, Jason-1, OSTM/Jason-2, and Jason-3 Version 5.0, PO.DAAC, CA, USA. Dataset accessed 25/04/2021 at http://dx.doi.org/10.5067/GMSLM-TJ150.

4   Data for Figure 1.2a are from Jiang et al. (2019); NOAA Ocean Acidification Data Stewardship Data Portal. Model data for Figure 1.2b are from the Geophysical Fluid Dynamics Laboratory (GFDL) ESM2M model (Dunne et al., 2012) in an RCP8.5 scenario, showing the change in SST over the 21st century.

6   The two models from which the cloud simulations are shown in Figure 1.3a,b are the GFDL model and the Hadley center model used in the IPCC AR5.

6   The record of the number of hurricanes per year in Figure 1.3c is from Vecchi and Knutson (2011); downloaded from www.gfdl.noaa.gov.

7   The sea ice images in Figure 1.4a are from the NASA/Goddard Space Flight Center scientific visualization studio, "Annual Arctic Sea Ice Minimum 1979–2019 with Area Graph"; the figure emulates the NASA animations. The sea ice area time series is from the National Snow and Ice Data Center (NSIDC; nsidc.org/data/g02135.html), as described by Fetterer et al. (2017). The data were obtained from Ian Eisenman: they were post-processed to fill in the missing sea ice area satellite observations at the pole and to linearly interpolate the missing data at 12/1987 and 1/1988; see Eisenman et al. (2014).

7   The mountain glacier length records in Figure 1.4b are plotted using the World Glacier Monitoring Service (WGMS) database (World Glacier Monitoring Service, 2020; Zemp et al., 2020).

8   Figures 1.5a,b are based on the analysis of a long present-day control run of the Community Earth System Model, available as part of the large ensemble project (CESM-LE; Kay et al., 2015).

8   Figure 1.5c is based on an analysis of the large ensemble of the Community Earth System Model (CESM-LE; Kay et al., 2015).

8   The blue curve in Figure 1.5d is a time series of global fire area from the Global Fire Emissions Database, www.globalfiredata.org, described by Giglio et al. (2013) and analyzed by Andela et al. (2017). The data for the red curve, the western US fire area, are from Abatzoglou and Williams (2016).

CHAPTER 2: GREENHOUSE

11   Thanks to Camille Hankel for writing the first draft of this chapter and many of the codes that are used here!

11 The image on the chapter title page, "Strawberry greenhouse" by Joi Ito, is licensed under the Creative Commons Attribution 2.0 Generic license, downloaded from Wikimedia. See a brief discussion of the warming mechanism in a real greenhouse vs the atmospheric greenhouse effect at the end of section 2.1.2.

19 For a more quantitative discussion of the greenhouse effect explained via the increase in the emission height with increasing CO_2 discussed in section 2.1.3, see Held and Soden (2000) or Pierrehumbert (2010), section 6.6. The emission height and its increase with increased CO_2 concentration are both a function of wavelength. For a CO_2 doubling the height increase can be significant in the central absorbing bands of CO_2, and can even reach the stratosphere, where the temperature increases with height and the argument presented in Figure 2.4 is not valid anymore. However, this is compensated by the stratospheric cooling (section 3.4) and, overall, the argument presented holds and the increase in emission height contributes dominantly to the greenhouse effect (Dufresne et al., 2020).

24 The radiative forcing definition follows chapter 8 in the fifth IPCC Assessment Report (Stocker et al., 2013), where it is also noted that one normally uses the "stratospherically adjusted radiative forcing," which is calculated allowing stratospheric temperatures to readjust to radiative equilibrium while holding other climate components fixed.

24 For an explanation of the logarithmic dependence of radiative forcing on CO_2 concentration, see Pierrehumbert (2010).

24 A quantitative estimate of the radiative forcing by different greenhouse gases and their CO_2-equivalent mixing ratio as a function of time since 1979 is given by Butler and Montzka (2016), updated in 2020 at www.esrl.noaa.gov/gmd/aggi/.

26 For more on the Clausius-Clapeyron equation, see Emanuel (1994) section 4.4.14, pp. 116–17.

26 If the water vapor feedback is even stronger, it can cause the overall outgoing LW radiation to weaken as the temperature is increasing. This can lead to a runaway greenhouse effect, as may have occurred on Venus (Ingersoll, 1969; Pierrehumbert, 2010).

27 That water vapor doubles climate sensitivity was already noted by Manabe and Wetherald (1967) and confirmed by recent climate models, together with the idea that the large-scale relative humidity is unchanged in CO_2-induced warming, to a first approximation.

CHAPTER 3: TEMPERATURE

31 The image on the chapter title page is of a thermometer made by Anders Celsius using his temperature scale, on display in the Gustavianum Museum, Uppsala, Sweden. It was taken by Dr. Derek Muller, the creator of the Veritasium science and engineering video site, www.veritasium.com, and YouTube channel. Thanks!

31 Thanks to high school student Jonathan Vardi for preparing some observations and model output and writing several codes that are used in this chapter.

32 The estimated past temperature anomaly record in Figure 3.1 is from Mann et al. (2008) and was downloaded from http://www.meteo.psu.edu/holocene/public_html/supplements/Multiproxy Means07/. The RCP projections are from the Geophysical Fluid Dynamics Laboratory (GFDL) CM3 climate model (Donner et al., 2011).

32 For the hockey-stick curve, see Mann et al. (1999).

33 The RCP8.5 projection used to draw Figure 3.2 is from the GFDL model.

34 The depth of penetration of surface temperature variations into the solid land can be shown to vary with the square root of the timescale of the surface temperature signal by solving the heat diffusion equation with a prescribed periodic temperature signal at the surface.

36 Equations 3.3 are from Held et al. (2010).

43 In a warmer climate the high-latitude atmospheric temperature inversion over land may be eliminated due to the greenhouse warming effect of low clouds that are formed by condensation of moist air coming from the ocean, leading to an enhanced surface warming and polar amplification; see Cronin and Tziperman (2015).

44 The results in Figure 3.7 are from the GFDL CM3 model (Donner et al., 2011).

44 An analysis of ocean heat storage during the recent "hiatus" period is given by, e.g., Balmaseda et al. (2013).

45 The results in Figure 3.8 are from the GFDL CM3 model (Donner et al., 2011).

47 For more on the stratospheric heat budget, see Mlynczak et al. (1999). That the simple three-layer model shows the stratospheric cooling to cause the mean stratospheric temperature to be lower than the mean tropospheric temperature should not be taken as a robust prediction but rather as a result of the simplifying model assumptions.

48 Stratospheric cooling as a response to CO_2 increase was predicted by Manabe and Wetherald (1967); see also the discussion of this important paper in Archer and Pierrehumbert (2011).

49 The workshop problem on "hiatus" via fitting of a linear trend was suggested by Jonathan Gilligan following the nice blog post at tamino.wordpress.com/2014/01/30/global-temperature-the-post-1998-surprise/.

CHAPTER 4: SEA LEVEL

53 Thanks to Xiaoting Yang for writing the first draft of this chapter and many of the codes that are used here!

54 Sea level data for blue and cyan curves in Figure 4.1 are from Jevrejeva et al. (2008). Data for the red and orange curves are from the Global Mean Sea Level Trend from Integrated Multi-Mission Ocean Altimeters TOPEX/Poseidon, Jason-1, OSTM/Jason-2, and Jason-3 Version 5.0 PO.DAAC, CA, USA. Data set accessed 25/04/2021 at http://dx.doi.org/10.5067/GMSLM-TJ150. Satellite data are calculated using a different baseline and were shifted to fit the Jevrejeva curve in the overlap period.

55 The data for Figure 4.2a are from the Goddard Space Flight Center, GSFC, 2017, described as: "Global Mean Sea Level Trend from Integrated Multi-Mission Ocean Altimeters TOPEX/Poseidon, Jason-1, OSTM/Jason-2 Version 4.2 PO.DAAC, CA, USA." Data set accessed 21/07/2020 at https://doi.org/10.5067/GMSLM-TJ150; see details in Beckley et al. (2017). Data for Figure 4.2b were digitized from Figure 13.11a in the sea level chapter of the 2013 report of the Intergovernmental Panel on Climate Change (IPCC), Church et al. (2013).

55 Revised, careful definitions of sea level–related terms are given by Gregory et al. (2019).

56 The estimates throughout the chapter of GMSL change due to warming at different ocean depths, due to different ice sheets, or due to mountain glaciers, water storage, and so on are from the IPCC report, Church et al. (2013).

60 Panel a in the figure in Box 4.1 uses data from Zachos et al. (2001), downloaded from NOAA's National Centers for Environmental Information (NCEI) Paleoclimatology Program, www.ncdc.noaa.gov. The plotted isotopic composition is a complex function of temperature, ice volume over land, and other factors and cannot be taken literally as representing temperature.

60 Panel b in the figure in Box 4.1 uses data from Medina-Elizalde et al. (2008), downloaded from NOAA's NCEI Paleoclimatology Program, www.ncdc.noaa.gov.

62 The mass balance of 41 monitored glaciers is discussed by the National Snow and Ice Data Center, https://nsidc.org/cryosphere/sotc/glacier_balance.html.

63   The data for Figure 4.3 are the same as used for Figure 4.1. The figure itself is modeled after Figure 2 of Jevrejeva et al. (2008), which shows acceleration rather than the rate of increase shown here.

64   SLP changes and the resulting projected sea level signal are discussed on p. 1193 in Church et al. (2013).

70   For more on Venice land subsidence and sea level rise, see Bock et al. (2012).

70   The natural processes that can lead to periodic decline or buildup of coastal landscapes via sediment transport are discussed by, e.g., Auerbach et al. (2015) and Wilson and Goodbred (2015).

71   The term *fingerprint* in section 4.2.3 has been superseded by *barystatic-GRD fingerprint* with GRD standing for "mass, Gravity, Rotation, and (land) Deformation." See Table 8 in Gregory et al. (2019).

72   Figure 4.7 is drawn from a calculation done by Xiaoting Yang as part of a final project in a sea level course taught by J. Mitrovica.

72   For the gravitational fingerprint of melting ice sheets, see, e.g., Mitrovica et al. (2018).

73   The workshop question on the relative contribution of added heat to sea level rise due to melting land ice vs expanding seawater is from section 4.4 of Griffies and Greatbatch (2012).

CHAPTER 5: OCEAN ACIDIFICATION

77   The image on the chapter title page, "Coral reef ecosystem at Palmyra Atoll National Wildlife Refuge, Palmyra Atoll, central equatorial Pacific Ocean" by the U.S. Fish & Wildlife Service—Pacific Region (photo credit: Jim Maragos/U.S. Fish and Wildlife Service), is licensed under the Creative Commons Attribution 2.0 Generic license, downloaded from Wikimedia.

77   For a global carbon budget estimate identifying the fate of emitted anthropogenic CO_2 and in particular the fractions absorbed by the ocean and the land biosphere, see Friedlingstein et al. (2019).

78   Data for Figure 5.1 are from Jiang et al. (2019), downloaded from the NOAA Ocean Acidification Data Stewardship Data Portal via ftp.

78   Olfactory effects of acidification are discussed by, e.g., Cripps et al. (2011) and Dixson et al. (2010), and effects on retinal function are examined by Chung et al. (2014).

80   The results in Figures 5.2, 5.3, 5.4, and 5.5 were calculated using mocsy 2.0 Fortran 95 routines to model ocean carbonate system thermodynamics by James Orr, LSCE/IPSL, CEA-CNRS-UVSQ, Gif-sur-Yvette, France, and Jean-Marie Epitalon, Geoscientific Programming Services, Toulouse, France, and are available on GitHub using their python interface. The mocsy package is described by Orr and Epitalon (2015).

91   The mechanism behind the 100 ppm drop in atmospheric CO_2 during the LGM is still not well understood. It has been suggested that, beyond the change in solubility with temperature, the productivity of ocean biology in the Southern Ocean may have played a role (Siegenthaler and Wenk, 1984; Knox and McElroy, 1984; Sarmiento and Toggweiler, 1984), although a picture consistent with available biogeochemical proxy observations has not been formulated yet. See Archer et al. (2000) for a description of some of the challenges involved; see Toggweiler (1999) and Gildor et al. (2002) for examples of attempts to address these challenges.

93   The discussion of the long-term decline of anthropogenic CO_2 in section 5.3.3 follows Archer (2005).

CHAPTER 6: OCEAN CIRCULATION

99   The image on the chapter title page, "Earliest known map of the Gulf Stream" (source: https://www.loc.gov/resource/g9112g.ct000136/) is in the public domain under the terms of Title 17, Chapter 1, Section 105 of the U.S. Code, downloaded from Wikimedia.

100   That the location of the atmospheric jet stream may be a factor in the warmth of Europe relative to similar latitudes in eastern North America is pointed out by Seager et al. (2002).

101   It has been suggested that most of the transport of AMOC can return to the surface adiabatically in the Southern Ocean rather than by diapycnal mixing (Wolfe and Cessi, 2011). It seems that explaining the robustly observed exponential vertical density profile in the ocean does require some diapycnal mixing, possibly in ocean margins near rough topography (Miller et al., 2020).

103   The figure in Box 6.1 is drawn from data of the World Ocean Circulation Experiment (WOCE). The density contours shown are of potential density at $[26.8, 27.3, 27.6] + 1000 \, \text{kg/m}^3$. To be more precise, water mass spreading mostly occurs over neutral density surfaces (McDougall, 1987).

103   For an analysis of ship observations suggesting a decline in AMOC, see Bryden et al. (2005), although aliasing by the seasonal cycle should be taken into account when interpreting this record, as seen in the workshop for this chapter (Peter Huybers, personal communication).

103   The RAPID observational campaign program is described by Rayner et al. (2011).

104   The RAPID data in Figure 6.1 were downloaded from https://rapid.ac.uk/rapidmoc/rapid_data /datadl.php. Data from the RAPID AMOC monitoring project is funded by the Natural Environment Research Council and are freely available from www.rapid.ac.uk/rapidmoc.

105   The RCP8.5 results for AMOC in Figures 6.2, 6.3, and 6.8 are from the CMIP5 results of the Geophysical Fluid Dynamics Laboratory (GFDL) model (Delworth et al., 2006).

106   For an attempt to justify two-dimensional (latitude-depth) models of AMOC, see, e.g., Wright and Stocker (1991).

106   The box model analysis here follows Marotzke's (2000) elegant simplification of the original Stommel (1961) box model, first introduced in Marotzke (1990).

113   For a stability analysis of an ocean general circulation model and a box model suggesting that the ocean circulation might be close to a point of abrupt collapse (bifurcation point, or tipping point), see Tziperman et al. (1994b). For an intercomparison of the response of multiple ocean circulation models to hysteresis forcing, showing a Stommel model–like response, see Rahmstorf et al. (2005).

114   The schematic Figure 6.7 is motivated by Gnanadesikan (1999); the figure and discussion are based on Youngs et al. (2020). The asymmetry in the outline of the warm water pool between the two hemispheres hints at the role of ocean eddies in the Southern Ocean, which lead to the sloped isopycnals there. A study of the robustness of the predictions of box models of increasing complexity is given by Thual and McWilliams (1992). An application demonstrating the usefulness of the Stommel model is demonstrated by Marotzke (2000). A hysteresis in a coupled ocean-atmosphere climate model is examined by, e.g., Li et al. (2013).

115   Abrupt changes (tipping points) and hysteresis (irreversibility) as a function of climate forcing including CO_2 changes were studied for Arctic sea ice (Eisenman and Wettlaufer, 2009), wintertime Arctic atmospheric convection and clouds (Abbot and Tziperman, 2008, 2009), subtropical stratocumulus clouds (Schneider et al., 2019), ice sheet dynamics (Schoof, 2007a; Robel et al., 2013), ice ages (Gildor and Tziperman, 2000), and more.

115   The model output used in Figure 6.8 is from the GFDL ESM2M model (Dunne et al., 2012).

116 For studies that examine the origin of the observed "North Atlantic warming hole," see, e.g., Piecuch et al. (2017) and Chemke et al. (2020).

117 Understanding vs simulation in climate science is nicely discussed by Held (2005).

CHAPTER 7: CLOUDS

121 The image on the chapter title page, "Puffy Rolling Clouds WTR-CL-14" (© www.shadowmeld. com), is licensed under the Creative Commons Attribution-Share Alike 4.0 International license, downloaded from Wikimedia.

121 Thanks to Minmin Fu for writing the first draft of this chapter and many of the codes that are used here!

122 The "Charney report" is Charney et al. (1979).

124 For details on CRE calculation, which involves (among other things) separating fast adjustments from slow ones and feedbacks that depend on the surface temperature warming from those that do not, see Stocker et al. (2013).

126 The lapse rate within the troposphere is more precisely determined by an equilibrium of both convection and radiation rather than moist convection alone. In the stratosphere, adiabatic expansion and cooling is supplemented by a stronger radiative heating due to SW absorption by the ozone layer, leading to a warming with height.

127 Regarding MSE conservation: The precise statement is that equivalent potential temperature is conserved under adiabatic moist convection. MSE is conserved approximately in adiabatic motion under the added assumption of a hydrostatic balance.

CHAPTER 8: HURRICANES

141 The image on the chapter title page, "Dramatic Views of Hurricane Florence from the International Space Station from 9/12," posted to Flickr by NASA/Goddard, is licensed under the Creative Commons Attribution 2.0 Generic license, downloaded from Wikimedia.

141 Thanks to high school student Jonathan Vardi for preparing observations and model output and writing several codes that are used in this chapter.

145 For more details on the mechanisms by which vertical shear disrupts hurricanes, see, e.g., Frank and Ritchie (2001) and Riemer et al. (2010).

145 The results in Figure 8.2 are based on the ERA5 reanalysis product of the European Center for Mid-range Weather Forecasting (ECMWF). The relation between wind shear in the North Atlantic MDR during El Niño and La Niña events is analyzed, e.g., by Zhu et al. (2012).

146 The derivation of potential intensity follows Kerry Emanuel's web page; see also Emanuel (1986).

148 El Niño's irregularity, which limits its predictability, is suggested to be a result of either a nonlinear chaotic behavior driven by the seasonal cycle (Tziperman et al., 1994a; Jin et al., 1994) or stochastic-like atmospheric weather forcing (Kleeman and Moore, 1997; Penland and Sardeshmukh, 1995).

150 The efficiency of converting heat to kinetic energy is different, of course, from that of an ideal Carnot engine; see Bister et al. (2011).

153 The strengthening of hurricanes frequently occurs when a particular upper level potential vorticity disturbance interacts with a low level developing hurricane; see, e.g., Montgomery and Farrell (1993). This is an example of the more general concept of rapid storm development via what's known as "transient amplification" (Farrell, 1988; Farrell and Ioannou, 1996).

153 The analysis of the correlation between PDI and SST in the main development area in the North Atlantic during the hurricane season (Figure 8.5) follows Emanuel (2005). The

data used here to calculate the PDI are from the AOML/NOAA HURDAT2 data set, https://www.aoml.noaa.gov/hrd/hurdat/Data_Storm.html.

154 The analysis in Figure 8.6 follows Kossin et al. (2020), and the data are from the supplementary information to that paper. Thanks to high school student Veer Gadodia for doing the analysis and helping to prepare the plot.

154 The concept of type I and type II errors has been used in the context of observed and predicted hurricane strength by Knutson et al. (2019, 2020).

155 The workshop problem calculating the fraction of major hurricanes is based on Figure 2 of Kossin et al. (2020).

CHAPTER 9: ARCTIC SEA ICE

157 The image on the chapter title page, "Ice Cruising in the Arctic" by Gary Bembridge from London, UK, is licensed under the Creative Commons Attribution 2.0 Generic license and downloaded from Wikimedia.

158 The sea ice area and extent time series data in Figure 9.1 used to calculate the observed trend in Figures 9.7 and 9.8 are from the National Snow and Ice Data Center (NSIDC; http://nsidc.org/data/g02135.html), as described by Fetterer et al. (2017). The Arctic data were obtained from Ian Eisenman: they were post-processed to fill in the missing sea ice area satellite observations at the pole and to linearly interpolate the missing data at 12/1987 and 1/1988; see Eisenman et al. (2014). The Antarctic data were downloaded directly from the NSIDC.

159 Sea ice concentration data for 1979 and 2018 in Figure 9.2 are from NSIDC, nsidc.org (Walsh et al., 2019).

160 For a possible role for sea ice in past abrupt climate change, see Gildor and Tziperman (2003).

160 Sea ice age and thickness time series in Figures 9.3, 9.4, and 9.5 are from the NSIDC (nsidc.org). Figure 9.3 is drawn following Tschudi et al. (2019b). Thickness data are from the output of the Pan-Arctic Ice-Ocean Modeling and Assimilation System (PIOMAS), downloaded from the Polar Science Center, psc.apl.uw.edu. Sea ice age data were derived from remotely sensed sea ice motion, and sea ice extent data were downloaded from nsidc.org/data/nsidc-0611; see Tschudi et al. (2019a).

161 Sea ice formation is described and illustrated in nsidc.org/cryosphere/seaice/characteristics /formation.html.

165 A summary of factors that played a role in the observed Arctic sea ice retreat is provided by Serreze et al. (2007).

166 The climate change detection analysis in section 9.2, and in particular Figures 9.7 and 9.8, follows Vinnikov et al. (1999). The figures are based on a time series from a long present-day control run of the Community Earth System Model (CESM), available as part of the large ensemble (LE) project (Kay et al., 2015).

168 For how winter-time atmospheric convection (section 7.2) and clouds over the Arctic Ocean can lead to a greenhouse feedback that may help eliminate winter sea ice, see Abbot and Tziperman (2008).

CHAPTER 10: GREENLAND AND ANTARCTICA

171 The image on the chapter title page, "Early Melt on the Greenland Ice Sheet," taken on June 15, 2016, by the Advanced Land Imager (ALI) on NASA's Earth Observing-1 satellite just inland from the coast of southwestern Greenland, is licensed under the Creative Commons Attribution 2.0 Generic license and downloaded from Wikimedia.

171 Thanks to Wanying Kang for writing the first draft of this chapter and many of the codes that are used here!

172 Possible evidence for ice over Greenland as early as 18 Myr ago based on ice-rafted debris is discussed by Thiede et al. (2011), while much more recent exposure and vegetation events of Greenland over the past few million years, based on sub-glacial sediments, are studied by Christ et al. (2021).

174 For how the temperature-precipitation feedback can play a role in ice age dynamics, see Källén et al. (1979) and Gildor and Tziperman (2000).

175 The data for Figure 10.1 are from the Geophysical Fluid Dynamics Laboratory (GFDL) ESM2G model (Dunne et al., 2012) runs piControl and RCP8.5.

176 Figure 10.2 reproduces Figure 3 from Oerlemans (1991).

178 For a review of crevasse observations, formation mechanisms, and consequences, see Colgan et al. (2016).

178 A mechanism that can lead to rift propagation in ice shelves is studied in, e.g., Lipovsky (2018).

180 For the role of hydrofracturing in the Larson B ice shelf collapse, see Robel and Banwell (2019).

180 Thanks to Alex Robel for Figure 10.4!

182 For more on MISI, see Weertman (1974) and Schoof (2007b). Among the feedbacks that may stabilize MISI on a retrograde slope are, e.g., sea level response (Gomez et al., 2010) and 3D effects (Gudmundsson et al., 2012).

183 A large melt water flux into the base of an ice sheet may organize into channels and drain efficiently rather than form a lubricating layer and accelerate the flow; see Schoof (2010).

187 Data for Figure 10.5 are from Velicogna et al. (2020), downloaded from https://www.ess.uci.edu /~velicogna/.

188 Figure 10.6 of the surface mass balance of Greenland and Antarctica in the RCP8.5 scenario is calculated using the Community Earth System Model large ensemble (CESM-LE; Kay et al., 2015). It is motivated by the analysis of the projected Antarctica SMB by Previdi and Polvani (2016), who also concluded that the signal due to climate change will emerge from the envelope of natural variability toward the middle of the 21st century.

CHAPTER 11: MOUNTAIN GLACIERS

193 The image on the chapter title page is composed of two images of the Muir Glacier (in Alaska) from the Glacier Photograph Collection, Boulder, Colorado, National Snow and Ice Data Center, downloaded from nsidc.org/data/g00472. The top image was taken in 1941 by William Osgood Field; the bottom, in 2004 by Bruce F. Molnia.

194 An updated mountain glacier mass balance using satellite data and its contribution to global sea level is presented by Hugonnet et al. (2021).

194 That 1 K of warming is equivalent to a decrease of 25% in the accumulation rate is stated by Oerlemans (2005).

194 For more on PDD used for SMB modeling, see Reeh (1991).

195 Figures 11.1, 11.2, and 11.3 are plotted using the World Glacier Monitoring Service (WGMS) database (World Glacier Monitoring Service, 2020; Zemp et al., 2020). All records that started before year 1960 whose glacier length variations relative to 1960 are within the range of ± 3 km and that had at least 10 points in the record were used.

195 An excellent discussion of tropical glacier retreat is found in the RealClimate.org blog entry by Pierrehumbert (2005).

196 Figure 11.2 follows the analysis of Oerlemans (2005).

196　An example of stable or increasing mass of mountain glaciers is given by Bolch et al. (2012) in the analysis of decreasing trends in most Himalayan glaciers vs a possible mass gain in the Karakoram range.

197　Sources for the plant carbon dates mentioned: Quelccaya ice cap in Peru: Thompson et al. (2006); Baffin Island: Pendleton et al. (2019); Svalbard: Miller et al. (2017); Ellesmere Island: La Farge et al. (2013).

198　The discussion and analysis of the relation between glacier extent and temperature in section 11.2 roughly follows Oerlemans (2005).

201　The globally and annually averaged surface temperature time series in Figure 11.5a is based on the HadCRUT4 data set of the UK Met Office Hadley Centre (Morice et al., 2012), downloaded from www.metoffice.gov.uk/hadobs/hadcrut4/data/current/ensemble_series_format.html.

203　Figure 11.6 is based on the Quelccaya ice cores published by Thompson et al. (2013) and Thompson et al. (2017), downloaded from the NOAA World Data Service for Paleoclimatology, Paleo Data Search Web Service, www.ncdc.noaa.gov/paleo-search/.

204　The shallow ice approximation also assumes that vertical shear in ice velocity dominates other shear terms. The ice thickness and flow solutions in Figures 11.7 and 11.8 were obtained by numerically solving the 1D shallow ice approximation. The solver used is essentially the 2D SIA Matlab code written by Ed Bueler, professor of mathematics (applied), Dept. of Mathematics & Statistics, University of Alaska Fairbanks. The Matlab code accompanied the course notes on "Numerical modeling of glaciers, ice sheets, and ice shelves," International Summer School in Glaciology, McCarthy Alaska, June 2016 (Bueler, 2021). The model code was downloaded from: https://github.com/bueler/ and converted from 2D and 1D, and from Matlab to python for making this plot.

CHAPTER 12: DROUGHTS AND PRECIPITATION

211　The image on the chapter title page, "Dry ground in the Sonoran Desert, Sonora, Mexico" by Tomas Castelazo, is licensed under the Creative Commons Attribution 3.0 Unported license, downloaded from Wikimedia.

211　Thanks to high school student Jonathan Vardi for preparing some observations and model data sets and writing several codes that are used in this chapter.

214　For more on the PDSI, see https://climatedataguide.ucar.edu/climate-data/palmer-drought-severity-index-pdsi.

215　The PDSI data drawn in Figure 12.1 are from the North American Drought Atlas (Cook et al., 2010).

216　Intensification of desert conditions by adiabatic compression due to flow on downward sloping isentropic surfaces was suggested to be forced by atmospheric heating due to monsoons, for example. The heating creates a Rossby wave circulation pattern to the west, which leads to winds down isentropic surfaces, and the resulting subsidence leads to drying that was estimated to account for 50% of the desertification effect in some regions (Rodwell and Hoskins, 1996).

218　The IOD was identified by Saji et al. (1999) and suggested not to be an independent mode of variability by, e.g., Zhao and Nigam (2015), who found the sea surface temperature pattern to be most likely being driven by ENSO, although they allowed for the possibility that there may be an independent Indian Ocean *subsurface* dipole variability mode.

219　While the general idea of remote sea surface temperature patterns driving teleconnections that can force drought conditions (Figure 12.2) is robust, the details in the figure may deviate from those in observations of California droughts due to various model biases. The drivers of the 2011–2014 California drought were analyzed by Seager et al. (2015).

219 Figure 12.2 is based on an analysis of a long present-day control run of the Community Earth System Model, available as part of the large ensemble project (CESM-LE; Kay et al., 2015). The composites are calculated as averages over the 5% driest California Januaries in the model output.

221 For an analysis of recent California droughts in the context of the past 1000 years, see Griffin and Anchukaitis (2014).

221 Tree-ring data for Figure 12.4 are from the NOAA Paleoclimatology Program archive, https://www.ncdc.noaa.gov/data-access/paleoclimatology-data/datasets/tree-ring.

222 The discussion of the medieval megadroughts in North America is based on Seager et al. (2007).

223 The two models in Figure 12.5 are from the Geophysical Fluid Dynamics Laboratory (GFDL; Delworth et al., 2006) and MIROC (Model for Interdisciplinary Research on Climate), developed by the University of Tokyo, NIES, and JAMSTEC (K-1 model developers, 2004; Nozawa et al., 2007).

226 For an analysis of the Hadley cell expansion and its possible mechanisms, see Lu et al. (2007).

226 Figure 12.6 is based on the output of CESM-LE (Kay et al., 2015). The height coordinate is approximated using the hypsometric equation from model pressure levels.

228 The atmospheric boundary layer moisture budget argument for the weakening of the Hadley cell is presented in Betts (1998) and Held and Soden (2006), where its limitations in explaining this weakening are discussed. A fuller recent analysis is given by, e.g., Chemke and Polvani (2019).

228 The discussion of "wet getting wetter, dry getting dryer" in section 12.6.2 follows Held and Soden (2006). The data used for Figure 12.7 are from CESM-LE (Kay et al., 2015).

229 The derivation of the relation (12.1) between the meridional moisture flux and $P - E$ is schematic, as it ignores the sphericity of the Earth, which would introduce some corrections to this equation, although this does not affect the precipitation scaling response derived using this equation. The derivation also ignores longitudinal variations; one can more generally consider the horizontal divergence of the moisture flux instead of its gradient in y; this again leads to the same scaling. See Held and Soden (2006).

232 The limitations of the "wet getting wetter, dry getting dryer" projection (section 12.6.2), in particular over land, are analyzed by, e.g., Chou et al. (2009) and Greve et al. (2014).

232 The analysis of precipitation extremes in section 12.6.3 closely follows O'Gorman and Schneider (2009a,b) and O'Gorman (2015).

232 Figure 12.9 is based on the output of CESM-LE (Kay et al., 2015).

236 Figure 12.10 reproduces Figure 6 of O'Gorman and Schneider (2009b).

236 The two scalings in Figure 12.10 may be used to deduce that the thermodynamic increase in precipitation rates calculated from equation (12.3) is expected to be slower than the increase in the column integral of moisture with temperature, but not necessarily slower than the increase in surface specific humidity. Depending on the weighting by omega and on the vertical profile of warming, the thermodynamic increase may even be faster than the increase in saturation moisture at the surface. Thanks to Paul O'Gorman for clarifying this.

238 The bucket model for soil moisture, including the various constants used, follows section 2E of Manabe (1969).

CHAPTER 13: HEAT WAVES

246 Figure 13.1 and the related discussion of the 2020 Siberian heat wave follow Ciavarella et al. (2021), who suggested that ACC contributed significantly to the probability of occurrence of this heat wave. The data for the figure were downloaded from the National Centers for Environmental Information, NOAA, https://www.ncdc.noaa.gov, station RSM00024266, Verhojansk.

247 For a discussion of the many indices used to define and characterize heat waves and their trends in the United States, see Smith et al. (2013).

247 Catastrophic avian mortality due to heat waves is discussed by McKechnie and Wolf (2010).

248 Much of the analysis in the heat waves chapter, including in Figures 13.2, 13.5, and 13.6, roughly follows Lau and Nath (2012) and Chang and Wallace (1987). The model output used is from the large ensemble of the Community Earth System Model (CESM-LE; Kay et al., 2015).

251 An analysis of European heat wave events shows that adiabatic compression is a dominant factor in leading to the hot conditions and that it is significantly larger than the enhanced radiative cooling due to the clear sky conditions that develop within the high pressure center (Zschenderlein et al., 2019).

253 Heat stress danger is normally quantified by measures that are more general than the wet bulb temperature in order to take into account additional relevant factors to temperature and relative humidity; see, e.g., Sherwood (2018), whose figure of the WBT is emulated by Figure 13.3. The importance of both temperature and relative humidity in causing mortality is analyzed by Mora et al. (2017).

CHAPTER 14: FOREST FIRES

261 The image on the chapter title page is described as follows: "During the first week of June 2009, Sustainable Resource Alberta burned nearly 8,000 hectares of forest in Western Alberta, just east of the Saskatchewan River Crossing on the Icefields Parkway and Highway 11." Photo by Cameron Strandberg (Rocky Mountain House, Alberta, Canada); licensed under the Creative Commons Attribution 2.0 Generic license, downloaded from Wikimedia.

262 Carbon emissions by fires during 1997–2016 are estimated by Werf et al. (2017).

262 The effect of tree species on differences in boreal fire dynamics between North America and Eurasia is discussed by Rogers et al. (2015).

262 For more on fire as a natural ecological process and on evolution of fire-adaptive plant strategies, see McLauchlan et al. (2020) and Lamont et al. (2019).

263 For an analysis of human effects on fire ignition in the United States, see Balch et al. (2017).

263 Fire deficit in the western United States is estimated from a 3000 yr proxy record by Marlon et al. (2012) and is evaluated for more recent decades using a model by Parks et al. (2015).

263 A discussion of how fire suppression management may increase the sensitivity to climate change is in Williams et al. (2019).

263 The effects of fire suppression in boreal forests are discussed by Johnson et al. (1998) and Johnson et al. (2001).

263 The non-monotonic relation between population density and fires is analyzed by Bistinas et al. (2013).

263 The episodic nature of fires over the past two millennia was observed in lake sediments from the Siskiyou Mountains, OR (Colombaroli and Gavin, 2010).

264 The correlation of VPD and fires is examined by Williams et al. (2015).

266 The time series for area burnt, VPD, and VPD due to ACC shown in Figure 14.1 and used in the calculations there are from Abatzoglou and Williams (2016), where the authors use many more fire indices to obtain a more robust estimate of the fire area attributed to ACC. Section 14.2 follows the analysis of these authors.

268 Figure 14.2 is based on the results of the Community Earth System Model large ensemble project (CESM-LE; Kay et al., 2015).

269 The role of the frequency of wet days and a comparison with the effects of VPD are examined by Holden et al. (2018) and Williams et al. (2019). The analysis of future projections based on VPD vs frequency of wet days was generously shared by Park Williams in private communication.

270　Forests vs savanna as alternate stable states, and the role of fires in maintaining these, are discussed by Staver et al. (2011).

270　The role of the Indian Ocean SST in leading to drought conditions over southeast Australia is examined by Ummenhofer et al. (2009).

270　The analysis of the connection between NINO3.4, the IOD index, SAM, and rain over the Murray-Darling Basin in section 14.3 is inspired by, follows, and uses the same data as King et al. (2020). This study uses another approach, not clustering, to also demonstrate the effect of natural variability modes on the Australian droughts during 2017–2019.

270　The time series of NINO3.4 in Figure 14.3 is from the NOAA site psl.noaa.gov/gcos_wgsp /Timeseries/Data/nino34.long.data; that of the IOD is from the same site, psl.noaa.gov /gcos_wgsp/Timeseries/Data/dmi.had.long.data; the monthly SAM time series is described by Marshall (2003) and is from www.nerc-bas.ac.uk/icd/gjma/sam.html; the monthly data averaged over the Murray-Darling Basin (Australia) is from the Bureau of Meteorology website, www.bom.gov.au/cgi-bin/climate/change/timeseries.cgi.

272　The effects of SAM on precipitation over Australia, and the mechanisms responsible for such effects in different seasons, are discussed by Hendon et al. (2007). Understanding droughts in the area of interest for the analysis presented is complicated by the fact that none of the factors discussed (IOD, SAM, ENSO) explains a large fraction of the variance of precipitation variability there.

272　The attribution of the Australian fires of 2019–2020 to ACC is analyzed by van Oldenborgh et al. (2021).

274　The data for Figures 14.5a,b were digitized from Figure 2 of Jolly et al. (2015). Figure 14.5c is a time series of global fire area downloaded from the Global Fire Emissions Database, www.globalfiredata.org, described by Giglio et al. (2013), and analyzed by Andela et al. (2017).

274　Global fire trends and the anthropogenic effects leading to the decline in Figure 14.5c are analyzed by Andela et al. (2017).

274　The discussion of future projections in section 14.4 follows Moritz et al. (2012).

276　Greening trends, as measured by a satellite-observed leaf area index (LAI), and their causes are discussed by Zhu et al. (2016).

276　Studies of the complex interaction between bark beetles and forest fires are reviewed by Hicke et al. (2012).

276　Question 4 follows Abatzoglou and Williams (2016), also followed in section 14.2.

BIBLIOGRAPHY

Abatzoglou, J. T. and Williams, A. P. (2016). Impact of anthropogenic climate change on wildfire across western US forests. *Proceedings of the National Academy of Sciences*, 113(42):11770–11775.

Abbot, D. S. and Tziperman, E. (2008). Sea ice, high latitude convection, and equable climates. *Geophysical Research Letters*, 35:L03702.

Abbot, D. S. and Tziperman, E. (2009). Controls on the activation and strength of a high latitude convective-cloud feedback. *Journal of the Atmospheric Sciences*, 66:519–529.

Andela, N., Morton, D. C., Giglio, L., Chen, Y., van der Werf, G. R., Kasibhatla, P. S., DeFries, R. S., Collatz, G. J., Hantson, S., Kloster, S., Bachelet, D., Forrest, M., Lasslop, G., Li, F., Mangeon, S., Melton, J. R., Yue, C., and Randerson, J. T. (2017). A human-driven decline in global burned area. *Science*, 356(6345):1356–1362.

Archer, D. (2005). Fate of fossil fuel CO_2 in geologic time. *Journal of Geophysical Research: Oceans*, 110(C9).

Archer, D. and Pierrehumbert, R. (2011). *The warming papers: The scientific foundation for the climate change forecast*. John Wiley & Sons.

Archer, D., Winguth, A., Lea, D., and Mahowald, N. (2000). What caused the glacial/interglacial atmospheric pCO_2 cycles? *Reviews of Geophysics*, 38(2):159–189.

Auerbach, L., Goodbred Jr., S., Mondal, D., Wilson, C., Ahmed, K., Roy, K., Steckler, M., Small, C., Gilligan, J., and Ackerly, B. (2015). Flood risk of natural and embanked landscapes on the Ganges-Brahmaputra tidal delta plain. *Nature Climate Change*, 5(2):153–157.

Balch, J. K., Bradley, B. A., Abatzoglou, J. T., Nagy, R. C., Fusco, E. J., and Mahood, A. L. (2017). Human-started wildfires expand the fire niche across the United States. *Proceedings of the National Academy of Sciences*, 114(11):2946–2951.

Balmaseda, M. A., Trenberth, K. E., and Källén, E. (2013). Distinctive climate signals in reanalysis of global ocean heat content. *Geophysical Research Letters*, 40(9):1754–1759.

Beckley, B. D., Callahan, P. S., Hancock III, D., Mitchum, G., and Ray, R. (2017). On the "Cal-Mode" correction to TOPEX satellite altimetry and its effect on the global mean sea level time series. *Journal of Geophysical Research: Oceans*, 122(11):8371–8384.

Betts, A. K. (1998). Climate-convection feedbacks: Some further issues: An editorial comment. *Climatic Change*, 39(1):35–38.

Bister, M., Renno, N., Pauluis, O., and Emanuel, K. (2011). Comment on Makarieva et al., "A critique of some modern applications of the Carnot heat engine concept: The dissipative heat engine cannot exist." *Proceedings of the Royal Society A: Mathematical, Physical and Engineering Sciences*, 467(2125): 1–6.

Bistinas, I., Oom, D., Sá, A. C., Harrison, S. P., Prentice, I. C., and Pereira, J. M. (2013). Relationships between human population density and burned area at continental and global scales. *PLoS One*, 8(12):e81188.

Bock, Y., Wdowinski, S., Ferretti, A., Novali, F., and Fumagalli, A. (2012). Recent subsidence of the Venice Lagoon from continuous GPS and interferometric synthetic aperture RADAR. *Geochemistry, Geophysics, Geosystems*, 13(3).

Bolch, T., Kulkarni, A., Kaab, A., Huggel, C., Paul, F., Cogley, J. G., Frey, H., Kargel, J. S., Fujita, K., Scheel, M., Bajracharya, S., and Stoffel, M. (2012). The state and fate of Himalayan glaciers. *Science*, 336(6079):310–314.

Bryden, H., Longworth, H., and Cunningham, S. (2005). Slowing of the Atlantic meridional overturning circulation at 25 degrees N. *Nature*, 438(7068):655–657.

Bueler, E. (2021). Chapter 8: Numerical modeling of ice sheets, streams, and shelves. In Fowler, A. and Ng, F., editors, *Glaciers and ice sheets in the climate system: The 2016 Karthaus Summer School Lecture Notes*. Springer.

Butler, J. H. and Montzka, S. A. (2016). The NOAA annual greenhouse gas index (AGGI). *NOAA Earth System Research Laboratory*, 58. Updated in 2020 at www.esrl.noaa.gov/gmd/aggi/.

Chang, F.-C. and Wallace, J. M. (1987). Meteorological conditions during heat waves and droughts in the United States Great Plains. *Monthly Weather Review*, 115(7):1253–1269.

Charney, J. G., Arakawa, A., Baker, D. J., Bolin, B., Dickinson, R. E., Goody, R. M., Leith, C. E., Stommel, H. M., and Wunsch, C. I. (1979). *Carbon dioxide and climate: A scientific assessment*. National Academy of Sciences.

Chemke, R. and Polvani, L. M. (2019). Exploiting the abrupt $4 \times CO_2$ scenario to elucidate tropical expansion mechanisms. *Journal of Climate*, 32(3):859–875.

Chemke, R., Zanna, L., and Polvani, L. M. (2020). Identifying a human signal in the North Atlantic warming hole. *Nature Communications*, 11(1):1–7.

Chou, C., Neelin, J. D., Chen, C.-A., and Tu, J.-Y. (2009). Evaluating the "rich-get-richer" mechanism in tropical precipitation change under global warming. *Journal of Climate*, 22(8):1982–2005.

Christ, A. J., Bierman, P. R., Schaefer, J. M., Dahl-Jensen, D., Steffensen, J. P., Corbett, L. B., Peteet, D. M., Thomas, E. K., Steig, E. J., Rittenour, T. M., Tison, J.-L., Blard, P.-J., Perdrial, N., Dethier, D. P., Lini, A., Hidy, A. J., Caffee, M. W., and Southon, J. (2021). A multimillion-year-old record of Greenland vegetation and glacial history preserved in sediment beneath 1.4 km of ice at Camp Century. *Proceedings of the National Academy of Sciences*, 118(13).

Chung, W.-S., Marshall, N. J., Watson, S.-A., Munday, P. L., and Nilsson, G. E. (2014). Ocean acidification slows retinal function in a damselfish through interference with gabaa receptors. *Journal of Experimental Biology*, 217(3):323–326.

Church, J. A., Clark, P. U., Cazenave, A., Gregory, J. M., Jevrejeva, S., Levermann, A., Merrifield, M. A., Milne, G. A., Nerem, R. S., Nunn, P. D., Payne, A. J., Pfeffer, W. T., Stammer, D., and Unnikrishnan, A. S. (2013). Sea level change. In Stocker, T. F., Qin, D., Plattner, G.-K., Tignor, M., Allen, S. K.,

Boschung, J., Nauels, A., Xia, Y., Bex, B., and Midgley, B., editors, *Climate change 2013: The physical science basis. Contribution of Working Group I to the Fifth Assessment Report of the Intergovernmental Panel on Climate Change*. Cambridge University Press.

Ciavarella, A., Cotterill, D., Stott, P., Kew, S., Philip, S., van Oldenborgh, G., Skålevåg, A., Lorenz, P., Robin, Y., Otto, F., Hauser, M., Seneviratne, S., Lehner, F., and Zolina, O. (2021). Prolonged Siberian heat of 2020 almost impossible without human influence. *Climatic Change*, 166(1):1–18.

Colgan, W., Rajaram, H., Abdalati, W., McCutchan, C., Mottram, R., Moussavi, M. S., and Grigsby, S. (2016). Glacier crevasses: Observations, models, and mass balance implications. *Reviews of Geophysics*, 54(1):119–161.

Colombaroli, D. and Gavin, D. G. (2010). Highly episodic fire and erosion regime over the past 2,000 y in the Siskiyou Mountains, Oregon. *Proceedings of the National Academy of Sciences*, 107(44):18909–18914.

Cook, E. R., Seager, R., Heim Jr., R. R., Vose, R. S., Herweijer, C., and Woodhouse, C. (2010). Megadroughts in North America: Placing IPCC projections of hydroclimatic change in a long-term palaeoclimate context. *Journal of Quaternary Science*, 25(1):48–61.

Cripps, I. L., Munday, P. L., and McCormick, M. I. (2011). Ocean acidification affects prey detection by a predatory reef fish. *Plos One*, 6(7):e22736.

Cronin, T. W. and Tziperman, E. (2015). Low clouds suppress Arctic air formation and amplify high-latitude continental winter warming. *Proceedings of the National Academy of Sciences*, 112(37):11490–11495. 10.1073/pnas.1510937112.

Delworth, T. L., Broccoli, A. J., Rosati, A., Stouffer, R. J., Balaji, V., Beesley, J. A., Cooke, W. F., Dixon, K. W., Dunne, J., Dunne, K. A., Durachta, J. W., Findell, K. L., Ginoux, P., Gnanadesikan, A., Gordon, C. T., Griffies, S. M., Gudgel, R., Harrison, M. J., Held, I. M., Hemler, R. S., Horowitz, L. W., Klein, S. A., Knutson, T. R., Kushner, P. J., Langenhorst, A. R., Lee, H. C., Lin, S. J., Lu, J., Malyshev, S. L., Milly, P.C.D., Ramaswamy, V., Russell, J., Schwarzkopf, M. D., Shevliakova, E., Sirutis, J. J., Spelman, M. J., Stern, W. F., Winton, M., Wittenberg, A. T., Wyman, B., Zeng, F., and Zhang, R. (2006). GFDL's CM2 global coupled climate models. Part I: Formulation and simulation characteristics. *Journal of Climate*, 19(5):643–674.

Dixson, D. L., Munday, P. L., and Jones, G. P. (2010). Ocean acidification disrupts the innate ability of fish to detect predator olfactory cues. *Ecology Letters*, 13(1):68–75.

Donner, L. J., Wyman, B. L., Hemler, R. S., Horowitz, L. W., Ming, Y., Zhao, M., Golaz, J.-C., Ginoux, P., Lin, S.-J., Austin, M.D.S.J., Alaka, G., Cooke, W. F., Delworth, T. L., Freidenreich, S. M., Gordon, C. T., Griffies, S. M., Held, I. M., Hurlin, W. J., Klein, S. A., Knutson, T. R., Langenhorst, A. R., Lee, H.-C., Lin, Y., Magi, B. I., Malyshev, S. L., Milly, P.C.D., Naik, V., Nath, M. J., Pincus, R., Ploshay, J. J., Ramaswamy, V., Seman, C. J., Shevliakova, E., Sirutis, J. J., Stern, W. F., Stouffer, R. J., Wilson, R. J., Winton, M., Wittenberg, A. T., and Zeng, F. (2011). The dynamical core, physical parameterizations, and basic simulation characteristics of the atmospheric component AM3 of the GFDL global coupled model CM3. *Journal of Climate*, 24(13):3484–3519.

Dufresne, J.-L., Eymet, V., Crevoisier, C., and Grandpeix, J.-Y. (2020). Greenhouse effect: The relative contributions of emission height and total absorption. *Journal of Climate*, 33(9):3827–3844.

Dunne, J. P., John, J. G., Adcroft, A. J., Griffies, S. M., Hallberg, R. W., Shevliakova, E., Stouffer, R. J., Cooke, W., Dunne, K. A., Harrison, M. J., Krasting, J. P., Malyshev, S. L., Milly, P.C.D., Phillipps, P. J., Sentman, L. A., Samuels, B. L., Spelman, M. J., Winton, M., Wittenberg, A. T., and Zadeh, N. (2012). GFDL's ESM2 global coupled climate-carbon Earth system models. Part I: Physical formulation and baseline simulation characteristics. *Journal of Climate*, 25(19):6646–6665.

Eisenman, I., Meier, W. N., and Norris, J. R. (2014). A spurious jump in the satellite record: Has Antarctic sea ice expansion been overestimated? *Cryosphere*, 8(4):1289–1296.

Eisenman, I. and Wettlaufer, J. S. (2009). Nonlinear threshold behavior during the loss of Arctic sea ice. *Proceedings of the National Academy of Sciences, USA*, 106:28–32.

Emanuel, K. (2005). Increasing destructiveness of tropical cyclones over the past 30 years. *Nature*, 436(7051):686–688.

Emanuel, K. A. (1986). An air-sea interaction theory for tropical cyclones. Part I: Steady-state maintenance. *Journal of the Atmospheric Sciences*, 43(6):585–605.

Emanuel, K. A. (1994). *Atmospheric Convection*. Oxford University Press.

Farrell, B. (1988). Optimal excitation of neutral Rossby waves. *Journal of the Atmospheric Sciences*, 45:163–172.

Farrell, B. F. and Ioannou, P. J. (1996). Generalized stability theory part I: Autonomous operators. *Journal of the Atmospheric Sciences*, 53:2025–2040.

Fetterer, F., Knowles, K., Meier, W., Savoie, M., and Windnagel, A. (2017). Sea ice index, version 3. *National Snow and Ice Data Center, Boulder, CO*. Updated daily. http://nsidc.org/data/g02135.html.

Frank, W. M. and Ritchie, E. A. (2001). Effects of vertical wind shear on the intensity and structure of numerically simulated hurricanes. *Monthly Weather Review*, 129(9):2249–2269.

Friedlingstein, P., Jones, M. W., O'Sullivan, M., Andrew, R. M., Hauck, J., Peters, G. P., Peters, W., Pongratz, J., Sitch, S., Le Quéré, C., Bakker, D.C.E., Canadell, J. G., Ciais, P., Jackson, R. B., Anthoni, P., Barbero, L., Bastos, A., Bastrikov, V., Becker, M., Bopp, L., Buitenhuis, E., Chandra, N., Chevallier, F., Chini, L. P., Currie, K. I., Feely, R. A., Gehlen, M., Gilfillan, D., Gkritzalis, T., Goll, D. S., Gruber, N., Gutekunst, S., Harris, I., Haverd, V., Houghton, R. A., Hurtt, G., Ilyina, T., Jain, A. K., Joetzjer, E., Kaplan, J. O., Kato, E., Klein Goldewijk, K., Korsbakken, J. I., Landschützer, P., Lauvset, S. K., Lefèvre, N., Lenton, A., Lienert, S., Lombardozzi, D., Marland, G., McGuire, P. C., Melton, J. R., Metzl, N., Munro, D. R., Nabel, J.E.M.S., Nakaoka, S.-I., Neill, C., Omar, A. M., Ono, T., Peregon, A., Pierrot, D., Poulter, B., Rehder, G., Resplandy, L., Robertson, E., Rödenbeck, C., Séférian, R., Schwinger, J., Smith, N., Tans, P. P., Tian, H., Tilbrook, B., Tubiello, F. N., van der Werf, G. R., Wiltshire, A. J., and Zaehle, S. (2019). Global carbon budget 2019. *Earth System Science Data*, 11(4):1783–1838.

Giglio, L., Randerson, J. T., and Van Der Werf, G. R. (2013). Analysis of daily, monthly, and annual burned area using the fourth-generation global fire emissions database (GFED4). *Journal of Geophysical Research: Biogeosciences*, 118(1):317–328.

Gildor, H. and Tziperman, E. (2000). Sea ice as the glacial cycles climate switch: Role of seasonal and orbital forcing. *Paleoceanography*, 15:605–615.

Gildor, H. and Tziperman, E. (2003). Sea-ice switches and abrupt climate change. *Philosophical Transactions of the Royal Society of London Series A: Mathematical Physical and Engineering Sciences*, 361(1810):1935–1942.

Gildor, H., Tziperman, E., and Toggweiler, R. J. (2002). The sea-ice switch mechanism and glacial-interglacial CO_2 variations. *Global Biogeochemical Cycles*, 16:10.1029/ 2001GB001446.

Gnanadesikan, A. (1999). A simple predictive model for the structure of the oceanic pycnocline. *Science*, 283:2077–2079.

Gomez, N., Mitrovica, J. X., Huybers, P., and Clark, P. U. (2010). Sea level as a stabilizing factor for marine-ice-sheet grounding lines. *Nature Geoscience*, 3(12):850–853.

Gregory, J. M., Griffies, S. M., Hughes, C. W., Lowe, J. A., Church, J. A., Fukimori, I., Gomez, N., Kopp, R. E., Landerer, F., Le Cozannet, G., Ponte, R. M., Stammer, D., Tamisiea, M. E., and van de Wal, R.S.W. (2019). Concepts and terminology for sea level: Mean, variability and change, both local and global. *Surveys in Geophysics*, 40(6):1251–1289.

Greve, P., Orlowsky, B., Mueller, B., Sheffield, J., Reichstein, M., and Seneviratne, S. I. (2014). Global assessment of trends in wetting and drying over land. *Nature Geoscience*, 7(10):716–721.

Griffies, S. M. and Greatbatch, R. J. (2012). Physical processes that impact the evolution of global mean sea level in ocean climate models. *Ocean Modelling*, 51:37–72.

Griffin, D. and Anchukaitis, K. J. (2014). How unusual is the 2012–2014 California drought? *Geophysical Research Letters*, 41(24):9017–9023.

Gudmundsson, H., Krug, J., Durand, G., Favier, L., and Gagliardini, O. (2012). The stability of grounding lines on retrograde slopes. *The Cryosphere*, 6(6):1497–1505.

Held, I. M. (2005). The gap between simulation and understanding in climate modeling. *Bulletin of the American Meteorological Society*, 86(11):1609–1614.

Held, I. M. and Soden, B. J. (2000). Water vapor feedback and global warming. *Annual Review of Energy and the Environment*, 25:441–475.

Held, I. M. and Soden, B. J. (2006). Robust responses of the hydrological cycle to global warming. *Journal of Climate*, 19(21):5686–5699.

Held, I. M., Winton, M., Takahashi, K., Delworth, T., Zeng, F., and Vallis, G. K. (2010). Probing the fast and slow components of global warming by returning abruptly to preindustrial forcing. *Journal of Climate*, 23(9):2418–2427.

Hendon, H. H., Thompson, D. W., and Wheeler, M. C. (2007). Australian rainfall and surface temperature variations associated with the Southern Hemisphere annular mode. *Journal of Climate*, 20(11):2452–2467.

Hicke, J. A., Johnson, M. C., Hayes, J. L., and Preisler, H. K. (2012). Effects of bark beetle-caused tree mortality on wildfire. *Forest Ecology and Management*, 271:81–90.

Holden, Z. A., Swanson, A., Luce, C. H., Jolly, W. M., Maneta, M., Oyler, J. W., Warren, D. A., Parsons, R., and Affleck, D. (2018). Decreasing fire season precipitation increased recent western US forest wildfire activity. *Proceedings of the National Academy of Sciences*, 115(36):E8349–E8357.

Hugonnet, R., McNabb, R., Berthier, E., Menounos, B., Nuth, C., Girod, L., Farinotti, D., Huss, M., Dussaillant, I., Brun, F., and Kaab, A. (2021). Accelerated global glacier mass loss in the early twenty-first century. *Nature*, 592(7856):726–731.

Ingersoll, A. P. (1969). The runaway greenhouse: A history of water on Venus. *Journal of the Atmospheric Sciences*, 26(6):1191–1198.

Jevrejeva, S., Moore, J., Grinsted, A., and Woodworth, P. (2008). Recent global sea level acceleration started over 200 years ago? *Geophysical Research Letters*, 35(8).

Jiang, L.-Q., Carter, B. R., Feely, R. A., Lauvset, S. K., and Olsen, A. (2019). Surface ocean pH and buffer capacity: Past, present and future. *Scientific Reports*, 9(1):1–11.

Jin, F.-F., Neelin, D., and Ghil, M. (1994). ENSO on the devil's staircase. *Science*, 264:70–72.

Johnson, E., Miyanishi, K., and Bridge, S. (2001). Wildfire regime in the boreal forest and the idea of suppression and fuel buildup. *Conservation Biology*, 15(6):1554–1557.

Johnson, E., Miyanishi, K., and Weir, J. (1998). Wildfires in the western Canadian boreal forest: Landscape patterns and ecosystem management. *Journal of Vegetation Science*, 9(4):603–610.

Jolly, W. M., Cochrane, M. A., Freeborn, P. H., Holden, Z. A., Brown, T. J., Williamson, G. J., and Bowman, D. M. (2015). Climate-induced variations in global wildfire danger from 1979 to 2013. *Nature Communications*, 6(1):1–11.

K-1 Model Developers (2004). K-1 coupled GCM (MIROC) description. Technical report no. 1, Center for Climate System Research, University of Tokyo, National Institute for Environmental Studies, and Frontier Research Center for Global Change.

Källén, E., Crafoord, C., and Ghil, M. (1979). Free oscillations in a climate model with ice-sheet dynamics. *Journal of the Atmospheric Sciences*, 36:2292–2303.

Kay, J. E., Deser, C., Phillips, A., Mai, A., Hannay, C., Strand, G., Arblaster, J. M., Bates, S., Danabasoglu, G., Edwards, J., Holland, M., Kushner, P., Lamarque, J.-F., Lawrence, D., Lindsay, K., Middleton, A.,

Munoz, E., Neale, R., Oleson, K., Polvani, L., and Vertenstein, M. (2015). The Community Earth System Model (CESM) large ensemble project: A community resource for studying climate change in the presence of internal climate variability. *Bulletin of the American Meteorological Society*, 96(8):1333–1349.

Keeling, C. D., Bacastow, R. B., Bainbridge, A. E., Ekdahl Jr., C. A., Guenther, P. R., Waterman, L. S., and Chin, J. F. (1976). Atmospheric carbon dioxide variations at Mauna Loa observatory, Hawaii. *Tellus*, 28(6):538–551.

King, A. D., Pitman, A. J., Henley, B. J., Ukkola, A. M., and Brown, J. R. (2020). The role of climate variability in Australian drought. *Nature Climate Change*, 10(3):177–179.

Kleeman, R. and Moore, A. M. (1997). A theory for the limitation of ENSO predictability due to stochastic atmospheric transients. *Journal of the Atmospheric Sciences*, 54:753–767.

Knox, F. and McElroy, M. B. (1984). Changes in atmospheric CO_2: Influence of the marine biota at high latitude. *Journal of Geophysical Research*, 89:4629–4637.

Knutson, T., Camargo, S. J., Chan, J. C., Emanuel, K., Ho, C.-H., Kossin, J., Mohapatra, M., Satoh, M., Sugi, M., Walsh, K., and Wu, L. (2019). Tropical cyclones and climate change assessment: Part I: Detection and attribution. *Bulletin of the American Meteorological Society*, 100(10):1987–2007.

Knutson, T., Camargo, S. J., Chan, J. C., Emanuel, K., Ho, C.-H., Kossin, J., Mohapatra, M., Satoh, M., Sugi, M., Walsh, K., and Wu, L. (2020). Tropical cyclones and climate change assessment: Part II: Projected response to anthropogenic warming. *Bulletin of the American Meteorological Society*, 101(3):E303–E322.

Kossin, J. P., Knapp, K. R., Olander, T. L., and Velden, C. S. (2020). Global increase in major tropical cyclone exceedance probability over the past four decades. *Proceedings of the National Academy of Sciences*, 117(22):11975–11980.

La Farge, C., Williams, K. H., and England, J. H. (2013). Regeneration of little ice age bryophytes emerging from a polar glacier with implications of totipotency in extreme environments. *Proceedings of the National Academy of Sciences*, 110(24):9839–9844.

Lamont, B. B., He, T., and Yan, Z. (2019). Evolutionary history of fire-stimulated resprouting, flowering, seed release and germination. *Biological Reviews*, 94(3):903–928.

Lau, N.-C. and Nath, M. J. (2012). A model study of heat waves over North America: Meteorological aspects and projections for the twenty-first century. *Journal of Climate*, 25(14):4761–4784.

Li, C., Notz, D., Tietsche, S., and Marotzke, J. (2013). The transient versus the equilibrium response of sea ice to global warming. *Journal of Climate*, 26(15):5624–5636.

Lipovsky, B. P. (2018). Ice shelf rift propagation and the mechanics of wave-induced fracture. *Journal of Geophysical Research: Oceans*, 123(6):4014–4033.

Lu, J., Vecchi, G. A., and Reichler, T. (2007). Expansion of the Hadley cell under global warming. *Geophysical Research Letters*, 34(6).

Manabe, S. (1969). Climate and the ocean circulation: I. The atmospheric circulation and the hydrology of the Earth's surface. *Monthly Weather Review*, 97(11):739–774.

Manabe, S. and Wetherald, R. (1967). Thermal equilibrium of atmosphere with a given distribution of relative humidity. *Journal of the Atmospheric Sciences*, 24(3):241–259.

Mann, M. E., Bradley, R. S., and Hughes, M. K. (1999). Northern hemisphere temperatures during the past millennium: Inferences, uncertainties, and limitations. *Geophysical Research Letters*, 26(6):759–762.

Mann, M. E., Zhang, Z., Hughes, M. K., Bradley, R. S., Miller, S. K., Rutherford, S., and Ni, F. (2008). Proxy-based reconstructions of hemispheric and global surface temperature variations over the past two millennia. *Proceedings of the National Academy of Sciences*, 105(36):13252–13257.

Marlon, J. R., Bartlein, P. J., Gavin, D. G., Long, C. J., Anderson, R. S., Briles, C. E., Brown, K. J., Colombaroli, D., Hallett, D. J., Power, M. J., Scharf, E. A., and Walsh, M. K. (2012). Long-term perspective on wildfires in the western USA. *Proceedings of the National Academy of Sciences*, 109(9):E535–E543.

Marotzke, J. (1990). *Instabilities and multiple equilibria of the thermohaline circulation*. PhD thesis, Berlin Instit Meereskunde, Kiel.

Marotzke, J. (2000). Abrupt climate change and thermohaline circulation: Mechanisms and predictability. *Proceedings of the National Academy of Sciences*, 97:1347–1350.

Marshall, G. J. (2003). Trends in the southern annular mode from observations and reanalyses. *Journal of Climate*, 16(24):4134–4143.

McDougall, T. J. (1987). Neutral surfaces. *Journal of Physical Oceanography*, 17:1950–1964.

McKechnie, A. E. and Wolf, B. O. (2010). Climate change increases the likelihood of catastrophic avian mortality events during extreme heat waves. *Biology Letters*, 6(2):253–256.

McLauchlan, K. K., Higuera, P. E., Miesel, J., Rogers, B. M., Schweitzer, J., Shuman, J. K., Tepley, A. J., Varner, J. M., Veblen, T. T., Adalsteinsson, S. A., Balch, J. K., Baker, P., Batllori, E., Bigio, E., Brando, P., Cattau, M., Chipman, M. L., Coen, J., Crandall, R., Daniels, L., Enright, N., Gross, W. S., Harvey, B. J., Hatten, J. A., Hermann, S., Hewitt, R. E., Kobziar, L. N., Landesmann, J. B., Loranty, M. M., Maezumi, Y., Mearns, L., Moritz, M., Myers, J. A., Pausas, J. G., Pellegrini, F. A., Platt, W. J., Roozeboom, J., Safford, H., Santos, F., Scheller, R. M., Sherriff, R. L., Smith, K. G., Smith, M. D., and Watts, A. C. (2020). Fire as a fundamental ecological process: Research advances and frontiers. *Journal of Ecology*, 108(5):2047–2069.

Medina-Elizalde, M., Lea, D. W., and Fantle, M. S. (2008). Implications of seawater Mg/Ca variability for Plio-Pleistocene tropical climate reconstruction. *Earth and Planetary Science Letters*, 269(3–4):584–594.

Miller, G. H., Landvik, J. Y., Lehman, S. J., and Southon, J. R. (2017). Episodic neoglacial snowline descent and glacier expansion on Svalbard reconstructed from the ^{14}C ages of ice-entombed plants. *Quaternary Science Reviews*, 155:67–78.

Miller, M., Yang, X., and Tziperman, E. (2020). Reconciling the observed mid-depth exponential ocean stratification with weak interior mixing and Southern Ocean dynamics via boundary-intensified mixing. *European Physical Journal Plus*, 135(375):1–15.

Mitrovica, J. X., Hay, C. C., Kopp, R. E., Harig, C., and Latychev, K. (2018). Quantifying the sensitivity of sea level change in coastal localities to the geometry of polar ice mass flux. *Journal of Climate*, 31(9):3701–3709.

Mlynczak, M. G., Mertens, C. J., Garcia, R. R., and Portmann, R. W. (1999). A detailed evaluation of the stratospheric heat budget: 2. Global radiation balance and diabatic circulations. *Journal of Geophysical Research: Atmospheres*, 104(D6):6039–6066.

Montgomery, M. T. and Farrell, B. F. (1993). Tropical cyclone formation. *Journal of the Atmospheric Sciences*, 50(2):285–310.

Mora, C., Dousset, B., Caldwell, I. R., Powell, F. E., Geronimo, R. C., Bielecki, C. R., Counsell, C. W., Dietrich, B. S., Johnston, E. T., Louis, L. V., Lucas, M. P., McKenzie, M. M., Shea, A. G., Tseng, H., Giambelluca, T., Leon, L. R., Hawkins, E., and Trauernicht, C. (2017). Global risk of deadly heat. *Nature Climate Change*, 7(7):501–506.

Morice, C. P., Kennedy, J. J., Rayner, N. A., and Jones, P. D. (2012). Quantifying uncertainties in global and regional temperature change using an ensemble of observational estimates: The HadCRUT4 data set. *Journal of Geophysical Research: Atmospheres*, 117(D8).

Moritz, M. A., Parisien, M.-A., Batllori, E., Krawchuk, M. A., Van Dorn, J., Ganz, D. J., and Hayhoe, K. (2012). Climate change and disruptions to global fire activity. *Ecosphere*, 3(6):1–22.

Nozawa, T., Nagashima, T., Ogura, T., Yokohata, T., Okada, N., and Shiogama, H. (2007). Climate change simulations with a coupled ocean-atmosphere GCM called the model for interdisciplinary research on climate: MIROC. *CGER's Supercomputer Monograph Report*, 12:80.

Oerlemans, J. (1991). The role of ice sheets in the Pleistocene climate. *Norsk Geologisk Tidsskrift*, 71(3):155–161.

Oerlemans, J. (2005). Extracting a climate signal from 169 glacier records. *Science*, 308(5722):675–677.

O'Gorman, P. A. (2015). Precipitation extremes under climate change. *Current Climate Change Reports*, 1(2):49–59.

O'Gorman, P. A. and Schneider, T. (2009a). The physical basis for increases in precipitation extremes in simulations of 21st-century climate change. *Proceedings of the National Academy of Sciences*, 106(35):14773–14777.

O'Gorman, P. A. and Schneider, T. (2009b). Scaling of precipitation extremes over a wide range of climates simulated with an idealized GCM. *Journal of Climate*, 22(21):5676–5685.

Orr, J. and Epitalon, J.-M. (2015). Improved routines to model the ocean carbonate system: mocsy 2.0. *Geoscientific Model Development*, 8(3).

Parks, S. A., Miller, C., Parisien, M.-A., Holsinger, L. M., Dobrowski, S. Z., and Abatzoglou, J. (2015). Wildland fire deficit and surplus in the western United States, 1984–2012. *Ecosphere*, 6(12):1–13.

Pendleton, S. L., Miller, G. H., Lifton, N., Lehman, S. J., Southon, J., Crump, S. E., and Anderson, R. S. (2019). Rapidly receding Arctic Canada glaciers revealing landscapes continuously ice-covered for more than 40,000 years. *Nature Communications*, 10(1):1–8.

Penland, C. and Sardeshmukh, P. D. (1995). The optimal-growth of tropical sea-surface temperature anomalies. *Journal of Climate*, 8(8):1999–2024.

Piecuch, C. G., Ponte, R. M., Little, C. M., Buckley, M. W., and Fukumori, I. (2017). Mechanisms underlying recent decadal changes in subpolar North Atlantic Ocean heat content. *Journal of Geophysical Research: Oceans*, 122(9):7181–7197.

Pierrehumbert, R. (2005). Tropical glacier retreat. RealClimate blog, http://www.realclimate.org/index.php/archives/2005/05/tropical-glacier-retreat/.

Pierrehumbert, R. T. (2010). *Principles of planetary climate*. Cambridge University Press.

Previdi, M. and Polvani, L. M. (2016). Anthropogenic impact on Antarctic surface mass balance, currently masked by natural variability, to emerge by mid-century. *Environmental Research Letters*, 11(9):094001.

Rahmstorf, S., Crucifix, M., Ganopolski, A., Goosse, H., Kamenkovich, I., Knutti, R., Lohmann, G., Marsh, R., Mysak, L. A., Wang, Z. M., and Weaver, A. J. (2005). Thermohaline circulation hysteresis: A model intercomparison. *Geophysical Research Letters*, 32(23).

Rayner, D., Hirschi, J. J.-M., Kanzow, T., Johns, W. E., Wright, P. G., Frajka-Williams, E., Bryden, H. L., Meinen, C. S., Baringer, M. O., Marotzke, J., Beal, L. M., and Cunningham, S. A. (2011). Monitoring the Atlantic meridional overturning circulation. *Deep Sea Research Part II: Topical Studies in Oceanography*, 58(17–18):1744–1753.

Reeh, N. (1991). Parameterization of melt rate and surface temperature in the Greenland ice sheet. *Polarforschung*, 59(3):113–128.

Riemer, M., Montgomery, M. T., and Nicholls, M. E. (2010). A new paradigm for intensity modification of tropical cyclones: Thermodynamic impact of vertical wind shear on the inflow layer. *Atmospheric Chemistry and Physics*, 10(7):3163–3188.

Robel, A., DeGiuli, E., Schoof, C., and Tziperman, E. (2013). Modes, scales and hysteresis of ice stream temporal variability. *Journal of Geophysical Research*, 118:925–936.

Robel, A. A. and Banwell, A. F. (2019). A speed limit on ice shelf collapse through hydrofracture. *Geophysical Research Letters*, 46(21):12092–12100.

Rodwell, M. J. and Hoskins, B. J. (1996). Monsoons and the dynamics of deserts. *Quarterly Journal of the Royal Meteorological Society*, 122(534):1385–1404.

Rogers, B. M., Soja, A. J., Goulden, M. L., and Randerson, J. T. (2015). Influence of tree species on continental differences in boreal fires and climate feedbacks. *Nature Geoscience*, 8(3):228–234.

Saji, N., Goswami, B., Vinayachandran, P., and Yamagata, T. (1999). A dipole mode in the tropical Indian Ocean. *Nature*, 401(6751):360–363.

Sarmiento, J. L. and Toggweiler, J. R. (1984). A new model for the role of the oceans in determining atmospheric pCO_2. *Nature*, 308:621–624.

Schneider, T., Kaul, C. M., and Pressel, K. G. (2019). Possible climate transitions from breakup of stratocumulus decks under greenhouse warming. *Nature Geoscience*, 12(3):163–167.

Schoof, C. (2007a). Ice sheet grounding line dynamics: Steady states, stability, and hysteresis. *Journal of Geophysical Research: Earth Surface*, 112(F3).

Schoof, C. (2007b). Marine ice-sheet dynamics. Part 1. The case of rapid sliding. *Journal of Fluid Mechanics*, 573:27.

Schoof, C. (2010). Ice-sheet acceleration driven by melt supply variability. *Nature*, 468(7325): 803–806.

Schwalm, C. R., Glendon, S., and Duffy, P. B. (2020). RCP8.5 tracks cumulative CO_2 emissions. *Proceedings of the National Academy of Sciences*, 117(33):19656–19657.

Seager, R., Battisti, D. S., Yin, J., Gordon, N., Naik, N., Clement, A. C., and Cane, M. A. (2002). Is the Gulf Stream responsible for Europe's mild winters? *Quarterly Journal of the Royal Meteorological Society*, 128:2563–2586.

Seager, R., Graham, N., Herweijer, C., Gordon, A. L., Kushnir, Y., and Cook, E. (2007). Blueprints for Medieval hydroclimate. *Quaternary Science Reviews*, 26(19–21):2322–2336.

Seager, R., Hoerling, M., Schubert, S., Wang, H., Lyon, B., Kumar, A., Nakamura, J., and Henderson, N. (2015). Causes of the 2011–14 California drought. *Journal of Climate*, 28(18):6997–7024.

Serreze, M. C., Holland, M. M., and Stroeve, J. (2007). Perspectives on the Arctic's shrinking sea-ice cover. *Science*, 315(5818):1533–1536.

Sherwood, S. C. (2018). How important is humidity in heat stress? *Journal of Geophysical Research: Atmospheres*, 123(21):11–808.

Siegenthaler, U. and Wenk, T. (1984). Rapid atmospheric CO_2 variations and ocean circulation. *Nature*, 308:624–625.

Smith, T. T., Zaitchik, B. F., and Gohlke, J. M. (2013). Heat waves in the United States: Definitions, patterns and trends. *Climatic Change*, 118(3–4):811–825.

Staver, A. C., Archibald, S., and Levin, S. A. (2011). The global extent and determinants of savanna and forest as alternative biome states. *Science*, 334(6053):230–232.

Stocker, T. F., Qin, D., Plattner, G.-K., Tignor, M., Allen, S. K., Boschung, J., Nauels, A., Xia, Y., Bex, V., and Midgley, P. M., editors (2013). IPCC, *Climate change 2013: The physical science basis. Contribution of Working Group I to the Fifth Assessment Report of the Intergovernmental Panel on Climate Change.* Cambridge University Press, pp. 1535.

Stommel, H. (1961). Thermohaline convection with two stable regimes of flow. *Tellus*, 13:224–230.

Thiede, J., Jessen, C., Knutz, P., Kuijpers, A., Mikkelsen, N., Nørgaard-Pedersen, N., and Spielhagen, R. F. (2011). Millions of years of Greenland ice sheet history recorded in ocean sediments. *Polarforschung*, 80(3):141–159.

Thompson, L., Davis, M., Mosley-Thompson, E., Beaudon, E., Porter, S., Kutuzov, S., Lin, P.-N., Mikhalenko, V., and Mountain, K. (2017). Impacts of recent warming and the 2015/2016 El Niño on tropical Peruvian ice fields. *Journal of Geophysical Research: Atmospheres*, 122(23): 12–688.

Thompson, L. G., Mosley-Thompson, E., Brecher, H., Davis, M., León, B., Les, D., Lin, P.-N., Mashiotta, T., and Mountain, K. (2006). Abrupt tropical climate change: Past and present. *Proceedings of the National Academy of Sciences*, 103(28):10536–10543.

Thompson, L. G., Mosley-Thompson, E., Davis, M., Zagorodnov, V., Howat, I., Mikhalenko, V., and Lin, P.-N. (2013). Annually resolved ice core records of tropical climate variability over the past 1800 years. *Science*, 340(6135):945–950.

Thoning, K. W., Tans, P. P., and Komhyr, W. D. (1989). Atmospheric carbon dioxide at Mauna Loa Observatory: 2. Analysis of the NOAA GMCC data, 1974–1985. *Journal of Geophysical Research: Atmospheres*, 94(D6):8549–8565.

Thual, O. and McWilliams, J. C. (1992). The catastrophe structure of thermohaline convection in a two-dimensional fluid model and a comparison with low-order box models. *Geophysical and Astrophysical Fluid Dynamics*, 64:67–95.

Toggweiler, J. R. (1999). Variation of atmospheric CO_2 by ventilation of the ocean's deepest water. *Paleoceanography*, 14:572–588.

Tschudi, M., Meier, W. N., Stewart, J. S., Fowler, C., and Maslanik, J. (2019a). EASE-Grid sea ice age, version 4. Technical report. NASA National Snow and Ice Data Center Distributed Active Archive Center, Boulder, CO.

Tschudi, M. A., Meier, W. N., and Stewart, J. S. (2019b). An enhancement to sea ice motion and age products. *The Cryosphere Discuss*. https://doi.org/10.5194/tc-2019-40.

Tziperman, E., Stone, L., Cane, M. A., and Jarosh, H. (1994a). El-Nino chaos: Overlapping of resonances between the seasonal cycle and the Pacific ocean-atmosphere oscillator. *Science*, 264(5155): 72–74.

Tziperman, E., Toggweiler, J. R., Feliks, Y., and Bryan, K. (1994b). Instability of the thermohaline circulation with respect to mixed boundary-conditions: Is it really a problem for realistic models? *Journal of Physical Oceanography*, 24(2):217–232.

Ummenhofer, C. C., England, M. H., McIntosh, P. C., Meyers, G. A., Pook, M. J., Risbey, J. S., Gupta, A. S., and Taschetto, A. S. (2009). What causes southeast Australia's worst droughts? *Geophysical Research Letters*, 36(4).

van Oldenborgh, G. J., Krikken, F., Lewis, S., Leach, N. J., Lehner, F., Saunders, K. R., van Weele, M., Haustein, K., Li, S., Wallom, D., Sparrow, S., Arrighi, J., Singh, R. K., van Aalst, M. K., Philip, S. Y., Vautard, R., and Otto, F.E.L. (2021). Attribution of the Australian bushfire risk to anthropogenic climate change. *Natural Hazards and Earth System Sciences*, 21(3):941–960.

Vecchi, G. A. and Knutson, T. R. (2011). Estimating annual numbers of Atlantic hurricanes missing from the HURDAT database (1878–1965) using ship track density. *Journal of Climate*, 24(6): 1736–1746.

Velicogna, I., Mohajerani, Y., Landerer, F., Mouginot, J., Noel, B., Rignot, E., Sutterley, T., van den Broeke, M., van Wessem, M., and Wiese, D. (2020). Continuity of ice sheet mass loss in Greenland and Antarctica from the GRACE and GRACE follow-on missions. *Geophysical Research Letters*, 47(8):e2020GL087291.

Vinnikov, K. Y., Robock, A., Stouffer, R. J., Walsh, J. E., Parkinson, C. L., Cavalieri, D. J., Mitchell, J. F., Garrett, D., and Zakharov, V. F. (1999). Global warming and Northern Hemisphere sea ice extent. *Science*, 286(5446):1934–1937.

Walsh, J. E., Chapman, W. L., Fetterer, F., and Stewart, S. (2019). Gridded monthly sea ice extent and concentration, 1850 onward, version 2. Technical report, National Snow and Ice Data Center, Boulder, CO.

Weertman, J. (1974). Stability of the junction of an ice sheet and an ice shelf. *Journal of Glaciology*, 13(67):3–11.

Werf, G. R., Randerson, J. T., Giglio, L., Leeuwen, T.T.v., Chen, Y., Rogers, B. M., Mu, M., Van Marle, M. J., Morton, D. C., Collatz, G. J., Yokelson, R. J., and Kasibhatla, P. S. (2017). Global fire emissions estimates during 1997–2016. *Earth System Science Data*, 9(2):697–720.

Williams, A. P., Abatzoglou, J. T., Gershunov, A., Guzman-Morales, J., Bishop, D. A., Balch, J. K., and Lettenmaier, D. P. (2019). Observed impacts of anthropogenic climate change on wildfire in California. *Earth's Future*, 7(8):892–910.

Williams, A. P., Seager, R., Macalady, A. K., Berkelhammer, M., Crimmins, M. A., Swetnam, T. W., Trugman, A. T., Buenning, N., Noone, D., McDowell, N. G., Hryniw, N., Mora, C. I., and Rahn, T. (2015). Correlations between components of the water balance and burned area reveal new insights for predicting forest fire area in the southwest United States. *International Journal of Wildland Fire*, 24(1):14–26.

Wilson, C. A. and Goodbred, J.S.L. (2015). Construction and maintenance of the Ganges-Brahmaputra-Meghna delta: Linking process, morphology, and stratigraphy. *Annual Review of Marine Science*, 7:67–88.

Wolfe, C. L. and Cessi, P. (2011). The adiabatic pole-to-pole overturning circulation. *Journal of Physical Oceanography*, 41(9):1795–1810.

World Glacier Monitoring Service. (2020). Fluctuations of glaciers database. Technical report, Zurich, Switzerland.

Wright, D. G. and Stocker, T. F. (1991). A zonally averaged ocean model for the thermohaline circulation. Part I: Model development and flow dynamics. *Journal of Physical Oceanography*, 21:1713–1724.

Youngs, M. K., Ferrari, R., and Flierl, G. R. (2020). Basin-width dependence of northern deep convection. *Geophysical Research Letters*, 47(15):e2020GL089135.

Zachos, J., Pagani, M., Sloan, L., Thomas, E., and Billups, K. (2001). Trends, rhythms, and aberrations in global climate 65 Ma to present. *Science*, 292(5517):686–693.

Zemp, M., Gärtner-Roer, I., Nussbaumer, S. U., Bannwart, J., Rastner, P., Paul, F., and Hoelzle, M. E. (2020). Global glacier change bulletin no. 3 (2016–2017). Technical report, ISC (WDS)/IUGG(IACS)/UNEP/UNESCO/WMO, World Glacier Monitoring Service (WGMS), Zurich, Switzerland. doi:10.5904/wgms-fog-2019-12.

Zhao, Y. and Nigam, S. (2015). The Indian Ocean dipole: A monopole in SST. *Journal of Climate*, 28(1):3–19.

Zhu, X., Saravanan, R., and Chang, P. (2012). Influence of mean flow on the ENSO–vertical wind shear relationship over the northern tropical Atlantic. *Journal of Climate*, 25(3):858–864.

Zhu, Z., Piao, S., Myneni, R. B., Huang, M., Zeng, Z., Canadell, J. G., Ciais, P., Sitch, S., Friedlingstein, P., Arneth, A., Cao, C., Cheng, L., Kato, E., Koven, C., Li, Y., Lian, X., Liu, Y., Liu, R., Mao, J., Pan, Y., Peng, S., Peñuelas, J., Poulter, B., Pugh, T.A.M., Stocker, B. D., Viovy, N., Wang, X., Wang, Y., Xiao, Z., Yang, H., Zaehle, S., and Zeng, N. (2016). Greening of the Earth and its drivers. *Nature Climate Change*, 6(8):791–795.

Zschenderlein, P., Fink, A. H., Pfahl, S., and Wernli, H. (2019). Processes determining heat waves across different European climates. *Quarterly Journal of the Royal Meteorological Society*, 145(724):2973–2989.

INDEX

Symbols

$\delta^{18}O$

ice ages	185
mountain glaciers	202
past warm climates	59

A

ablation	173
zone, ice sheets	174
zone, mountain glaciers	194
absorption	
lines	20
longwave, by CO_2	21
longwave, by H_2O	21
SW, melt ponds	176
SW, sea ice	161
absorptivity	14
ACC, *see* Anthropogenic Climate Change	
accumulation	173
zone, ice sheets	174
zone, mountain glaciers	194
adiabatic process	125
AMOC	283

clouds	125
drying, deserts	225
extreme precipitation	234
hurricane efficiency	150
lapse rate feedback	41
MSE conservation	126
subsidence, droughts	216
warming, heat waves	248
wet bulb temperature	252
aerosols	131
direct effects	133
indirect effects	133
African easterly waves	142
albedo	13
clouds	123
feedback	39
ice	176
land	13
melt ponds	176
polar amplification	39
snow	176
alkalinity	83
carbonate	85
AMO, *see* Atlantic multidecadal oscillation	

AMOC, *see* Atlantic meridional overturning circulation

Andes glaciers 195

anomaly
    fire season length 273
    glacier length 199
    sea level 54
    soil moisture 224
    surface temperature 2, 31, 200, 213, 216

Antarctic Bottom Water 102

Antarctic Intermediate Water 102

Anthropogenic Climate Change
    detection vs attribution 48

Archimedes' law 71

Arctic amplification, *see* polar amplification

Atlantic meridional overturning circulation 100
    abrupt climate change 160
    box model 106
    collapse consequences 115
    hysteresis 113
    multiple equilibria 113
    sea ice 160
    Stommel model 106
    streamfunction 105
    tipping point 112

Atlantic multidecadal oscillation 104
    droughts 217

attribution of ACC, *see* detection and attribution

B

basal
    friction 179
    lubrication 181
    melting 182
    temperature 183

bicarbonate ion 78

black body 13
    radiation 20

blocking
    heat waves 250

Boltzmann constant 20

boundary layer
    droughts 216
    Hadley cell expansion 227
    hurricanes 144

BP, *see* before present

bucket model for soil moisture 237
    equations 238

buffer effect 91

bulk coefficient
    evaporation 149, 238
    friction 146

C

calcium carbonate 78
    dissolution at high CO_2 90
    solubility product 80

calving 173, 177
    buoyancy force 178
    cliff instability 177
    hydrofracturing 178
    rift propagation 177
    slumping 178

carbon cycle 81

carbon dating 197

carbonate ion 78

carbonate system 81
    approximate solution 87
    buffer effect 91
    equations 85
    solution 86

carbonic acid 78

Carnot cycle 149

CCN, *see* cloud condensation nuclei

CDF, *see* cumulative distribution function

CFC 24

Clausius-Clapeyron 26
    Arctic lapse rate feedback 43
    atmospheric convection 125
    droughts 215
    Hadley cell expansion 227
    hurricanes 144, 148
    MSE conservation 127
    precipitation extremes 234
    sea level 61
    surface mass balance 174

tropical lapse rate feedback   42
VPD, fires   264
wet getting wetter   230
cliff instability   177
climate sensitivity
   equilibrium   36
   of glaciers   199
   transient   39
climatic water deficit   264
climatology
   droughts   218
   heat waves   250
   SST, hurricanes   145
cloud condensation nuclei   133
cloud radiative effect   134
   climate uncertainty   134
clouds
   cirrus   124
   climate uncertainty   137
   dissipation   133
   droplet size distribution   132
   feedbacks   134
   feedbacks, energy balance   135
   height   124
   high vs low   124
   LW emissivity   123
   microphysics   131
   mixed phase   123
   model disagreement   134
   mountain glaciers   196
   parameterization   122
   radiative effect   123
   stratiform   131
   stratocumulus   124
   subgrid scale   131
   SW albedo   123
   water vs ice   124
cluster analysis   271
CO_2
   absorption bands   20
   doubling   23, 32, 35, 38, 57, 134
   doubling, cloud feedbacks   136
   fertilization   275
   land sink   78
   ocean absorption   91
   ocean sink   78

   ocean solubility and warming   91
   solubility pump   91
   vs other greenhouse gases   24
CO_2-equivalent mixing ratio   24, 32
coastal flooding   67
composite analysis
   droughts   218
   El Niño, hurricanes   145
   heat waves   248
condensation
   clouds   126
   hurricanes   142
   precipitation extremes   234
   tropical lapse rate feedback   42
convection   126
   heat waves   249
   hurricanes   144
   lapse rate   19
   ocean, AMOC   100
   winter sea ice melting   168
Coriolis force
   AMOC   106
   droughts   215
   heat waves   248
   hurricanes   142
   sea level   67
CRE, *see* Cloud radiative effect
crevasses   177
cumulative distribution function
   extreme precipitation   233
   heat waves   257
CWD, *see* climatic water deficit

D

daily maximum temperature   247
Dansgaard-Oeschger events   160
desert belt   225
detection and attribution   48
   Antarctic sea ice   164
   droughts   219
   error types I and II   154
   forest fires   272
   Greenland and Antarctica   187
   heat waves   257
   hurricanes   152
   mountain glaciers   207

detection and attribution (*continued*)
    ocean acidification   94
    ocean circulation, AMOC   104
    precipitation extremes   237
    sea ice   165
    sea level, global   63
    sea level, regional   67
    stratospheric cooling   48
detection of ACC, *see* detection and attribution
DIC, *see* dissolved inorganic carbon
dimensional analysis   184
dissolved inorganic carbon   83
droughts
    agricultural   212
    Australia fires   269
    California, La Niña   218
    future projections   222
    hydrological   212
    megadroughts   221
    megadroughts, La Niña   222
    meteorological   212
    reconstruction   221
    Sahel   222
    socioeconomic   212
    Southwest United States   224
dry static energy   127
    droughts   216
    heat waves   249
Dust Bowl   251
dynamic vegetation models   273

E

El Niño   146
    droughts   217
    hurricanes   145
    Pliocene   61
El Niño–Southern Oscillation   146, 287
ELA, *see* equilibrium line altitude
elevation-desert effect   174
Ellesmere Island glaciers   198
emission height   18
emissivity   14
    clouds   123
    sea ice melting   168
    stratospheric cooling   46

energy balance   13
    atmosphere   14
    cloud feedbacks   134
    ice surface   183
    one-layer model   14
    stratospheric cooling   46
    surface   14
    two-layer model   15
ensemble model runs   253, 254, 267
ENSO, *see* El Niño–Southern Oscillation
Eocene   59
equation of state   58
equilibrium line altitude   204
    ice sheets   174
    mountain glaciers   195
error types I and II   154
evaporation
    droughts   224
    hurricanes   149
    ocean salinity   108
    potential   238
    soil resistance   238
evapotranspiration   214
    potential   214

F

fires
    area burnt   265
    area burnt versus VPD   265
    Australia   269
    Australia, projections   273
    climate factors   262
    deficit   263
    fire indices   265
    global projections   273
    human factors   263
    lightnings   262
    projections uncertainty   275
    season length   273
    western US   265
fossil fuels   82

G

geopotential height   247

geostrophy

   hurricanes 143

   sea level 69

geothermal heat flux 182

gigaton carbon 81

Glen's law 179

global mean sea level

   Greenland, Antarctica 61, 172

global mean surface

    temperature

   glacier length 201

   observed 31

   projected, RCP 31

global warming hiatus 33

global warming potential 24

GMSL, *see* global mean sea

   level

GRACE satellites 186

gradient balance 143

graupel 131

Great Plains 247

greenhouse gases

   absorption bands 20

   Doppler broadening 22

   global warming potential 24

   lifetime 24

   pressure broadening 21

   water vapor feedback 27

greening 275

grounding line 173, 179

   MISI 180

groundwater table 238

GtC, *see* gigaton carbon

Gulf Stream 67, 100

GWP, *see* global warming

   potential

H

Hadley cell 225

   desert belts 225

halocline 165

heat capacity 34, 37

   atmosphere 34

   land 35

   ocean 34

heat stress 251

heat waves

CDF 257

decadal statistics 253

PDF 254

projected statistics 255

projections 253

subsidence 248

surface winds 250

vs droughts 251

Henry's law 82

   buffer effect 91

heterogeneous nucleation 133

"hiatus" periods 33, 44

high pressure

   droughts 214

   heat waves 247

hockey-stick curve 32

Holocene 185

homogeneous nucleation 133

humidity, *see* specific humidity

hurricanes

   convection 144

   development 142

   energetics 145

   energy input 148

   eye wall 142

   observed increase 152

   power dissipation 146

   shear, ventilation 144

   warm upper ocean 148

hydrofracturing 178

hydrometeors 131

hydrostatic balance

   atmospheric pressure

    profile 129

   MSE conservation 284

   sea level 66

I

ice

   heat conductivity 183

   heat diffusivity 184

ice cores

   dust layers 202

   isotopic composition 202

   melt signal 203

   seasonal cycle 203

ice shelf 173

ice streams 173
   acceleration 181
   back-pressure 179
   retrograde slope 181
IN, *see* ice nuclei
Indian Ocean dipole 217
   Australia fires 269
   droughts 217
instability
   AMOC 110
   MISI 181
Intergovernmental Panel on
   Climate Change 12, 281
inversion 43, 252
IOD, *see* Indian Ocean dipole
IPCC, *see* Intergovernmental
   Panel on Climate Change
isostatic adjustment 71
isotherms
   droughts 216
isotopic composition
   ice cores 202

K

Kilimanjaro glaciers 195
kinetic energy
   hurricanes 149

L

La Niña 146
   droughts 218
   forest fires 271
   hurricanes 145
land, *see* soil
   albedo 13
   biosphere 78
   greening 275
   precipitation 231
   water storage, sea level 62
lapse rate
   dry adiabatic 127
   feedback, Arctic 43
   feedback, tropical 43
   greenhouse effect 17
   ice sheets 174

   moist adiabatic 127
   mountain glaciers 195
latent heat
   flux, droughts 248
   heat waves 250
   hurricanes 148
   lapse rate 42
   of condensation, evaporation 126
   of melting, ice sheets 183
   release 126
   sea ice 163
lateral friction 179
LCL, *see* lift condensation level
leaf area index 290
level of free convection 129
level of neutral buoyancy 129
LFC, *see* level of free
   convection
LHS, *see* left-hand side
lift condensation level 129
   precipitation extremes 234
Little Ice Age
   mountain glaciers 194
   sea level 54
log-normal distribution 132
longwave radiation 14
   Arctic amplification 41, 43
   energy balance 14, 15
   ice energy balance 183
   Planck feedback 40
   trapped by clouds 122
LW, *see* longwave radiation

M

main development area 145
marginal ice zone 157
marine ice sheet instability 180
marine-based ice sheet 174
MDR, *see* main development area
melt ponds
   ice sheets 176
   sea ice 161
meridional moisture flux 229
methane 24
microphysics 131
   extreme precipitation 235

MISI, *see* marine ice sheet
    instability
mixed layer   56, 102
    hurricanes   144
MOC, *see* meridional
    overturning circulation
moist adiabatic lapse rate   41
moist adiabatic temperature
    profile   41
moist static energy   126
    graphical solution   128
    lapse rate feedback   43
moulins   181
Mount Kenya glaciers   195
mountain glaciers
    accelerated recent retreat   206
    adjustment time   199
    adjustment to warming   202
    and temperature   198
    carbon dating the retreat   197
    climate sensitivity   199
    clouds   196
    dynamics of response   205
    equilibrium length   199
    front location   194
    global trends   197
    lapse rate   195
    length   194
    length records   197
    Little Ice Age   197
    retreat exposes plants   197
    runoff   194
    sea level   62, 194
    tropical   195
MSE, *see* moist static energy
Murray-Darling Basin   271

N

negative feedback
    Arctic amplification   41
    sea ice thickness   163
    tropical lapse rate   43
NINO3.4 index
    Australian droughts, fires   271
    hurricanes   147
nitrous oxide   24

North Atlantic Deep Water   102
    AMOC   100
    Arctic sea ice   165

O

ocean
    acidification   78
    circulation, AMOC   100
    cloud feedbacks   135
    CO_2 dissolution   29
    deep, warming   37
    delaying warming   39
    evaporation, hurricanes   142
    evaporation, SMB   174
    freezing   161
    ice sheet calving   178
    precipitation   213
    roles in global warming   116
    temperature, salinity   102
    upper, warming   37
    warming and sea level   56
    water density   57
    water masses   102
OLR, *see* outgoing longwave
    radiation
omega, saturation state   81
outgoing longwave radiation   18
    Arctic amplification   41
    clouds   134
    cooling, droughts   216
    cooling, heat waves   249
    emission height   18
    energy balance   14
    ice surface temperature   183
ozone
    layer   17, 45–46
    toxic, fires   262

P

Palmer Drought Severity
    Index   214
parameterization
    calving   178
    cloud microphysics   132
    clouds   122

parameterization (*continued*)
    extreme precipitation   235
    moist convection   131
    ocean mixing   106
past warm climates
    Eocene   59
    Pliocene   61
pCO_2 92
PDD, *see* positive degree days
PDF, *see* probability distribution
    function
PDI, *see* power dissipation index
PET, *see* potential
    evapotranspiration
pH   78
    as a function of CO_2 86
    response to CO_2 increase   90
photon energy   20
Planck constant   20
Planck feedback   40
Planck's Law   20
Pliocene   61
polar amplification   33, 39
    albedo   39
    lapse rate feedback   43
    Planck feedback   40
positive degree days   175, 195
positive feedback
    atmospheric convection   126
    clouds   136
    hurricanes   143
    ice sheet melt ponds   177
    MISI   181
    runaway greenhouse   280
    sea ice age, roughness   162
    sea ice albedo   161
    sea ice mobility   163
    water vapor   27
potential evapotranspiration   214
    droughts, bucket model   238
    fires   264
potential intensity   150
power dissipation index   146
precipitation
    clouds   133
    deficit and fires   269
    desert belts   225

    droughts   224
    dry getting drier   228
    efficiency, extremes   234
    enhancing fires   262
    extremes   231
    Hadley cell   225
    heat waves   248
    meridional moisture flux   229
    wet getting wetter   228
probability distribution function
    cloud droplet sizes   132
    droughts   220
    heat waves   254
    sea ice trends   165
prograde slope   181

Q

Quelccaya ice cap, Peru   198, 202

R

radiation
    longwave   14
    shortwave   14
    visible   14
radiative cooling
    climate sensitivity   35
    clouds   131
    droughts   216
    heatwaves   249
radiative effect
    clouds   123
radiative forcing   23, 34
    CO_2 vs other gases   24
    CO_2, logarithmic behavior   23
RAPID   103
RCP, *see* Representative
    Concentration Pathway
RCP8.5   24
    realism and motivation   xii
regression
    fires   265
    hurricanes   153
    sea ice   166
relative humidity   25, 126
    convection   130

droughts 215
evaporation, droughts 239
evaporation, hurricanes 149
fires 264
fixed in global warming 27
global precipitation
projections 230
heat stress 251
heat waves 249
Representative Concentration
Pathway 12, 24
retrograde slope 180
RH, *see* relative humidity
RHS, *see* right-hand side
Rossby waves 216
runaway greenhouse warming 280

S

Sahel droughts 222
salinity
AMOC 107
definition 102
North Atlantic 102
reaction constants 86
sea level 57
seawater density 107
SAM, *see* southern annular mode
Santa Ana winds 262
saturation specific humidity, *see*
Clausius-Clapeyron
saturation water vapor pressure 25
scale height 129
sea ice
age 160, 161
albedo feedback 161
Antarctic 164
area 159
concentration 159
decline trend 167
difficulty of attribution 164
extent 159
floes 159
frazil ice 161
grease ice 161
insulation feedback 163
loss, impacts 159

mobility 163
observed record 158
pancake ice 161
projections 168
roughness 161
thickness 160
variability PDF 166
winter melting 168
sea level
coastal erosion 70
global mean 55
gravitational fingerprint 72
high tide 67
iostatic adjustment 71
land water storage 62
mountain glaciers 62, 194
pressure loading 64
relative 55
rise, ocean warming 57
steric 55
storm surge 67
wind stress 65
sea surface temperature
AMOC collapse 115
forest fires 269
hurricanes 143
teleconnections, droughts 216
seawater
carbonate system 78
density, AMOC 106
heat capacity 37
thermal expansion 58
shallow ice approximation 206
shear
ice flow 179
wind, hurricanes 144
shortwave radiation 14
cloud albedo 122
enhanced, droughts 215
enhanced, heat waves 249
mountain glaciers 196
sea ice albedo 161
slumping, ice cliffs 178
SMB, *see* surface mass balance
soil
carbon cycle 81
droughts 215, 251

soil (*continued*)

evaporation	238
heat capacity	35
moisture	237, 238
moisture, heat waves	248
moisture, PDSI	214
moisture, projections	224
porosity	238
water content	238

solar constant	13
southern annular mode	269
southern oscillation	146
specific heat capacity	37
droughts	216
heat waves	249
ice	184
MSE	126
ocean water	37
specific humidity	
droughts	215
forest fires	264
heat waves	249
moist convection, clouds	125
orders of magnitude	123
precipitation extremes	234
saturation	125
saturation, Clausius-Clapeyron	26
saturation, hurricanes	148
soil moisture	238
wet bulb temperature	252
spectral radiance	20
speed of light	20
Stefan-Boltzmann constant	13
Stokes velocity	133
storm surge	67
stratospheric cooling	45
streamfunction	
AMOC	105
Hadley cell	225
subduction, carbon cycle	81
subsidence	
droughts	215
heat waves	248
land, sea level	70
supersaturation	131
surface mass balance	173, 203
ice sheets	176
mountain glaciers	195

PDD	176
sea level	61
surface temperature	31
cloud feedbacks	135
daily maximum	248
energy balance	13
extreme precipitation	233
glacier extent	200
heat waves	248
ice sheet	175, 183, 184
MSE conservation	126
soil, evaporation	238
Sverdrup	100
SW, *see* shortwave radiation	

T

teleconnections	
drought	217
forest fires	269
temperature-precipitation feedback	174
terminal velocity	133
The Day After Tomorrow	116
thermocline	56, 102
tipping point	
AMOC	106, 112
climate examples	115
MISI	181
total CO_2	83
toy models	114
transpiration	214
tree rings	
drought reconstruction	221
past fire reconstruction	262
turbulence	
around falling drops	133
cloud formation	126
entrainment	131
ocean mixing, AMOC	103, 106

U

updraft	
clouds	133
extreme precipitation	235
upwelling	61

V

vapor pressure deficit	264
volcanoes, carbon cycle	81
VPD, *see* vapor pressure deficit	

W

water vapor, *see* specific humidity	
water vapor feedback	26–27
West Antarctic ice sheet	172
wet bulb temperature	251
wind	
evaporation	238
evapotranspiration	214
fire indices	265
forest fires	262
high pressure, droughts	215
high pressure, heat waves	247
hurricane	141
hurricane, eye wall	142
low pressure, hurricane	142
moisture flux	229
ocean circulation	101
sea ice transport	163
shear, hurricanes	144
stress, bulk formula	66
stress, sea level	66
surface, heat waves	250
updrafts, precipitation	235

Y

yield stress	177